Darwin Retried

Darwin Retried

AN APPEAL TO REASON

Norman Macbeth

 A DELTA BOOK

A DELTA BOOK

Published by Dell Publishing Co., Inc.
1 Dag Hammarskjold Plaza
New York, New York 10017
Copyright © 1971 by Norman Macbeth
All rights reserved,
including the right to reproduce this book
or parts thereof in any form.
For information contact
Gambit, Incorporated.
DELTA ® TM 755118, Dell Publishing Co., Inc.
Published by arrangement with
Gambit, Incorporated
Boston, Massachusetts 02108
Printed in the United States of America
First Delta Printing—August 1973

Contents

Darwin Retried

When the centenary of *The Origin of Species* was celebrated in 1959, I was living in Switzerland as a semi-invalid largely retired from the practice of law. In an idle moment I picked up a volume of essays commemorating the event and read them straight through. This was the beginning of my biological studies. In the next three years I continued them in a random way by reading four paperbacks on evolution: Sir Julian Huxley, *Evolution in Action* (Mentor, 1957); John Maynard Smith, *The Theory of Evolution* (Penguin, 1958); Garrett Hardin, *Nature and Man's Fate* (Mentor, 1961); and Loren Eiseley, *Darwin's Century* (Doubleday Anchor, 1961).

As a result of this reading, and of keeping my eyes and ears open, a number of new things dawned on me. This was only to be expected in one whose biological education, meager to begin with, was already thirty years old, but the new findings were not of the expectable nature. Instead of filling the large gaps in my mind, they upset the little store of solid matter. Here are some of the major surprises.

1. The biologists of the German-speaking world had no feeling that Darwinism had finally solved the problem of evo-

lution. The Russians, under Lysenko, were even less convinced. The French had many doubts.

2. The four authors had no respect for such important tenets of classical Darwinism as the biogenetic law, survival of the fittest, and the struggle for existence.

3. There were no longer any trees showing descent.

4. The books stressed genetics and mathematics rather than the accomplishments of breeders and the marvels of adaptation. The men and the atmosphere were far removed from Darwin and his circle of naturalists. Professor Hardin referred to genetics as a "numbers game." [1]

5. The writers did not always agree with reason, or with each other, or even with themselves. Thus Hardin said (129): "We cannot, at present, tell a living X-sperm from a living Y-sperm," but a few lines later he said that the two differed "genetically as well as visually": thus implying that things could differ visually when no one could tell them apart. Sir Julian Huxley (35) spoke of an unhappy time "when we were still ignorant of the mechanism of heredity," forgetting that he had just said (24): "Unfortunately, the precise way genes act during development is still very imperfectly understood." Infallibility was so manifestly lacking that it seemed permissible for a layman to form his own opinions, especially since the disputes usually turned on assumptions, inferences, and extrapolations rather than on biological observations.

The next phase in my education took place over dinner tables. If conversation lagged, I asked friends whether they knew that Darwinism was going to pieces, that there was no struggle for existence, and that the scholars no longer spoke about the survival of the fittest. The responses were illuminating. They showed blind and universal faith in the doctrines learned many years earlier in college survey courses, and full conviction that the Scopes trial in Tennessee had laid all

doubts to rest in 1925.* My friends would not believe me
without documentation. The conversation became lively.

This led to the long and continuing third phase, in which I
read the professional literature as well as popularizations and
labored to put a coherent case on paper. I had to begin by
sorting out the premises, including evolution itself. Here
Eiseley, who is excellent on historical perspective, was very
useful. I am grateful to him for showing me that evolution
has two aspects: one large and relatively easy, the other
smaller and much more difficult. All biologists know this,
but it is often forgotten.[2]

The first aspect arose with the youthful sciences of geology
and paleontology, as the strata and the fossils were uncovered
and classified. This work showed that many plants and animals appeared for the first time in the strata that lay higher
and were therefore presumed to be younger, while others appeared only in the lower strata. From these observations it was
not difficult to infer that there had been changes in the course
of time, numerous species having been added or eliminated
since the beginning. This inference is the large and easy aspect of evolution. Eiseley shows that it was reasonably well
known by 1800.

The second aspect is the modus operandi, the how and
why. Assuming that there has been change, progress, or evolution, how did it occur? Answers began to be ventured almost as soon as the problem was defined. Erasmus Darwin,

---

* John Scopes, a high-school teacher, was indicted in Dayton, Tennessee,
for the crime of teaching evolution in defiance of a Tennessee statute for-
bidding such teaching in the public schools. William Jennings Bryan, former Secretary of State and Democratic presidential candidate, came down
to act as special prosecutor, while Clarence Darrow, a famous liberal lawyer
in Chicago, acted as chief counsel for the defense. The press gathered in
vast numbers. The legal proceedings were confused and inconclusive, but
the trial offered a wonderful opportunity for Darrow to ridicule Bryan's
fundamentalist views. The scientific problems received no sensible discussion.

grandfather of Charles, wrote extensively on the subject before his death in 1802. Jean Baptiste Lamarck, who died in 1829, propounded his much-derided suggestion that change comes about through acquired habits being passed on to later generations. Charles Darwin, when he came forward with *The Origin of Species* in 1859, was addressing himself to this latter aspect. Natural selection was his modus operandi, his answer to how and why.

Then there was the problem of a definition, since none of my four paperbacks gave a neat short summary of classical Darwinism. Some statement being necessary if there was to be careful discussion, I worked out the following:

> On the basis of data drawn from comparative anatomy, embryology, and the experience of breeders, classical Darwinism asserted that the progression from the early species to the later ones, as observed in the rocks, was a process of actual physical descent governed by natural selection through such agencies as the struggle for existence, survival of the fittest, sexual selection, and adaptation, all of which worked in small cumulative steps through vast periods of relatively undisturbed time. This had two logical corollaries: first, in the evolution of any structure or function, every intermediate stage must be of advantage to the species; second, natural selection tends to make each being only as perfect as, or slightly more perfect than, the other inhabitants of the same area, and does not produce absolute perfection.

I have submitted this formulation to several biologists and am sure that it is reasonably fair and accurate.[3] The components are, of course, familiar doctrine to everyone who has attended an American university and, if my samplings of opinion are reliable, they are accepted without question by almost every such person. The major doubters are the professional biologists.

When these preliminaries had been settled, I wrote a short article contending that classical Darwinism was dead. This was almost entirely based on the verbatim statements of eminent biologists, my part being mainly to select and arrange. I made it perfectly clear that I was not imitating William Jennings Bryan by attacking the large and easy aspect of evolution. I made it equally clear that I was not discussing genetics or neo-Darwinism. My thesis was simply that the professionals had moved away from classical Darwinism, but that no one had informed the public of what had happened. This, I believed, was important news for the American public.

My article was published in the *Yale Review*, together with a reply by Edward S. Deevey, Jr., professor of ecology at Yale. I found Professor Deevey's argument extremely hard to follow, but he seemed to be rejecting my thesis and advising me to read the works of Simpson, Lack, and Fisher. This I have done with great care, and the result has been that my confidence has increased. I still think that the news is correct and important, and that it should be made common knowledge. The purpose of this book is to make the news available to the public.

Courtroom experience during my career at the bar taught me to attach great weight to something that may seem trivial to persons not skilled in argumentation—the burden of proof. The proponents of a theory, in science or elsewhere, are obligated to support every link in the chain of reasoning, whereas a critic or skeptic may peck at any aspect of the theory, testing it for flaws. He is not obligated to set up any theory of his own or to offer any alternative explanations. He can be purely negative if he so desires. William Jennings Bryan forgot this in Tennessee, and was jockeyed into trying to defend fundamentalism, although this was not necessary to the matter in hand. The results were disastrous. They would have been equally disastrous for Clarence Darrow if he had tried to discharge the burden of proof for the other side. The winner in these matters is the skeptic who has no case to prove.

So let the following points be nailed down at once. I am not denying evolution in the large sense. I am not discussing genetics. I am not defending fundamentalism or propounding any other theory. I assert only that the mechanism of evolution suggested by Charles Darwin has been found inadequate by the professionals, and that they have moved on to other views and problems. In brief, classical Darwinism is no longer considered valid by qualified biologists.

Now a word as to my sources. The evolutionists are at present in a condition of remarkable harmony. There is quibbling about details, but something close to complete unanimity on general principles.[4] The leading exponents of the prevailing doctrine (the synthetic theory) are de Beer, Fisher, Ford, Haldane, Huxley, and Waddington in England, and Dobzhansky, Mayr, Muller, Simpson, Stebbins, and Wright in the United States.[5] Three of the authors of my original four paperbacks, Eiseley, Hardin, and Smith, are not ranked in this inner group; they are popularizers rather than authorities, but with the possible exception of Eiseley they are in complete accord with the synthetic theory.

I quote Simpson, Mayr, and Sir Julian Huxley more than any others, simply because they make the clearest and most vigorous statements of the points involved. I have tried to quote them as representatives of the fraternity rather than for any maverick opinions of their own (which they do not have in any case). As a result, the reader may be fairly confident that he is getting the consensus of the profession and that I am not setting up any illusory controversies or easily refuted straw men.

I have been rather surprised to discover that many biologists dispute the propriety of a purely skeptical position. They assert that the skeptic is obligated to provide a better theory than the one he attacks. Thus Professor Ernst Mayr of Harvard rules out admittedly valid objections on the ground that the objectors have not advanced a better suggestion.[6] I thought at first that this was a personal foible of Mayr's, but it has recurred in so many other places that it must be a widespread opinion.

I cannot take this view seriously. If a theory conflicts with the facts or with reason, it is entitled to no respect. As T. H. Huxley long ago remarked of men of science, ". . . there is not a single belief that it is not a bounden duty with them to hold with a light hand and to part with cheerfully, the moment it is really proved to be contrary to any fact, great or small." [7] Whether a better theory is offered is irrelevant.

I have been even more surprised to find that it is no longer considered proper for a scientist to approach his work with pure observation, avoiding any preconceived theories and even any working hypotheses. This would be pure induction, about which Ghiselin says ". . . a more pernicious fallacy could scarcely be enunciated." [8] The scientist, it is asserted, simply cannot proceed without having a number of assumptions (amounting to a theory) in his mind, and the best he can hope for is that he will be conscious of these assumptions.

Though far from persuaded that this is a correct view, I will not actively dissent if three modest qualifications are granted by the reader:

(a) The theory must be reasonable, not merely the only one or the best one.

(b) The scientist must hold his theory "with a light hand."

(c) The scientist must part with his theory cheerfully the moment it is proved to be contrary to any fact, great or small.

These qualifications are modest but they should not be accepted as a matter of course, as though they were self-evident and self-executing. We all know of instances where theories were *not* held with a light hand and were *not* parted with cheerfully. Many more such instances will be presented in the following chapters.

1. Hardin (1961), 133.
2. Simpson (1964), 10: "The fact—not theory—that evolution has occurred and the Darwinian theory as to how it has occurred have become so confused in popular opinion that the distinction must be stressed."
3. It is fortunate that I did not have to formulate the now dominant synthetic theory, since this seems to be an almost impossible task; Olson (1960), 525–527.
4. De Beer (1966), 30; Mayr (1963), 8.
5. Simpson (1949), 273.
6. Mayr (1942), 296.
7. T. H. Huxley (1893), 468–469.
8. Ghiselin (1969), 212; see also de Beer (1970), 32.

Darwinism is deeply indebted to comparative anatomy and embryology, but these are the working tools of the students and are not in themselves material for controversy. The biologists, with only an occasional exception, are in substantial harmony as to these disciplines. I want to make some reservations as to their implications, but will not otherwise dissent.

At the beginning of his major work *Animal Species and Evolution*, Professor Ernst Mayr of Harvard has a passage that sounded boastful to me when I first read it: "Genetics, morphology, biogeography, systematics, paleontology, embryology, physiology, ecology, and other branches of biology, all have illuminated some special aspect of evolution and have contributed to the total explanation where other special fields failed. In many branches of biology one can become a leader even though one's knowledge is essentially confined to an exceedingly limited area. This is unthinkable in evolutionary biology. A specialist can make valuable contributions to special aspects of the evolutionary theory, but only he who is well versed in most of the above-listed branches of biology can present a balanced picture of evolution as a whole. Whenever a narrow specialist has tried to develop a new theory of evolution, he has failed." [1] I have slowly been convinced that this is the plain truth and that there is no puffing in Mayr's words. The theorists must be, and are, extremely learned men. They may not always be correct in their reasoning, but no one can deny the breadth of their knowledge. By the same token, however, they are a very small band and the run-of-the-mill biologist can do little to test or verify them.

Embryology is mentioned in Mayr's list, while comparative anatomy is not. This is only an oversight. He and every other evolutionist have been trained in comparative anatomy. It is taken for granted, since morphology and paleontology could not exist without it. Nowadays one must also reckon with comparative physiology and ethology (behavior), both of which have grown into large separate disciplines.

The study of anatomy sets the stage for the study of evolution. It lays bare the resemblances and dissimilarities that gave rise to the whole subject. It shows, as nothing else can, the innumerable and wonderful variations that nature constructs on certain themes, such as the vertebrate skeleton. Particular parts suddenly swell and dominate the form: the upper lip in the elephant, the neck in the giraffe, one toe in the horse. Or an analogous structure is built up by different species in different ways, as in the wings of pterodactyls, bats, and birds. There is nothing better than comparative anatomy for opening the eyes and stirring the imagination. This is the sort of thing that Darwin found, on a small scale, among the Galápagos finches, and the effect on his thinking was enormous.[*]

Embryology was very close to Darwin's heart. In *The Origin of Species* he devoted twelve solid pages to it,[3] plus brief references at many other points. He was confident that his

[*] Professor Deevey of Yale calls comparative anatomy "an embalmed sort of scholasticism that was kept in the curriculum as a sop to the medical schools." I cannot understand this when the subject is indispensable for all zoologists, and especially for all paleontologists; but then the paleontologists are, in Deevey's eyes, "almost the sole custodians of a forgotten subject." [2] I find no one else taking such a position.

theory would clear up many embryological puzzles, and that embryology in turn would aid his theory in many ways. Thus he remarked: ". . . in the eyes of most naturalists, the structure of the embryo is even more important for classification than that of the adult. For the embryo is the animal in its less modified state; and in so far it reveals the structure of its progenitor. In two groups of animals, however much they may at present differ from each other in structure and habits, if they pass through the same or similar embryonic stages, we may feel assured that they have both descended from the same or nearly similar parents, and are therefore in that degree closely related." [4]

Ernst Haeckel (1834-1919), who carried the banner for Darwinism in Germany, extended this line of thought much further than Darwin did. He propounded what is known as the biogenetic law, which declared that the growth of the embryo was a recapitulation of the history of the species. The implication was that embryology would provide us with the lines of descent that are so conspicuously missing among adult forms.

There is no doubt that embryology furnishes occasional hints as to relationships. Here is an example from Smith: ". . . there is little resemblance in adult structure between flat-worms, annelids (segmented worms), and molluscs (snails, bi-valves, cephalopods). Yet some members of all these three phyla show the same pattern of cleavage of the fertilized egg into many cells, a pattern known as spiral cleavage. This pattern is so characteristic that it seems unlikely to have arisen independently more than once; it is therefore concluded that the three phyla are descended from a common ancestor in which early development followed this pattern, although this tells us little about the adult structure of this ancestor." [5] If the reader doubts whether such hints should be taken seriously and whether such reasoning from improbability is justified (see Chapter 10), he will be in good company. The biogenetic law has been popular in Germany, before as well as after Haeckel's time (it was not wholly original); but the English-speaking biologists were never willing to accept it.[6] Sir Gavin de Beer says that it is more misleading than helpful and should be rejected.[7] G. G. Simpson refers to it as "the overgeneralized and much abused aphorism of the nineteenth century." [8]

This does not mean that embryology is no longer important in evolutionary studies. It has value even in botany, where the layman is not accustomed to think of embryos.[9] But a century of experience shows that it is limited to furnishing hints and that it will not provide full explanations.[10] Thus it does much less than Haeckel asserted and a good deal less than Darwin hoped.

The important thing to realize is that embryology and comparative anatomy, although they are admirable disciplines and part of the evolutionist's equipment in any country, are not in themselves *solutions*. They show that certain animals and plants resemble each other, from which we infer (rightly or wrongly) that they are cousins; but they do not supply us with the information that the evolutionist so earnestly seeks, which is where the animals and plants came from. They do not give us what the scientists call the phylogenies, meaning the family trees or lines of descent. Since these phylogenies are precisely what the layman expects from the evolutionist, the failure to furnish them is a fact of gigantic importance. We will refer to it repeatedly.

Now let me dilate on the fact that the modern texts shy away from the charts of descent that were so dear to Haeckel and other early evolutionists. Not one of my four paperbacks includes a tree, although Hardin shows a couple of tangled thickets to demonstrate how hopeless the problem is.[11] This reticence is not because our modern authors are more ignorant than their ancestors; it is due to bitter experience with the frailty of all conjectures in this field. I will show two vulnerable points.

First, if a paleontologist portrays known fossil forms such as sharks, fishes, and amphibians in man's family tree, we soon perceive that many examples of these forms are still present, practically unchanged, although their former siblings are said to have worked up to human status. Thus one and the same ancient stock split into a group with astonishing plasticity and another group with almost total rigidity. This is very hard to swallow.[12] Therefore there was a tendency, while trees were still being used, to move the known forms out to the branches and reserve the trunk for malleable forms not yet discovered. The number of blank spaces was a visible and embarrassing sign of ignorance.

Second, the reader will become familiar (in Chapter 15) with a surprising line of thought developed by Robert Broom and Sir Julian Huxley to the effect that all specialized forms are dead ends because they can only evolve a little further in the same direction, remain unchanged, or die out. This applies to fossils as well as to living forms, since with the possible exception of man all known forms, extinct or extant, early or late, are already specialized. Therefore these forms, being ineligible as ancestors, must again be moved from the trunk of the tree to the branches. The result is that the tips are well populated while the trunk is shrouded in mist and mystery. We have the paradox that the remains found in the earth's crust are not those of our ancestors, while the bodies of our ancestors have not been preserved at all. We see forms that purport to be our cousins, but we have no idea of who our common grandparents were.

At this point the attentive reader will object that he has seen charts of descent with his own eyes, particularly those showing the progress of horses from the tiny *Eohippus* to the modern *Equus*. This is correct; he has seen precisely this picture on a number of occasions. But this leads to a story that is very embarrassing to the profession. I give it in Hardin's summary:

> . . . there was a time when the existing fossils of the horses seemed to indicate a straight-line evolution from small to large, from dog-like to horse-like, from animals with simple grinding teeth to animals with the complicated cusps of the modern horse. It looked straight-line—like the links of a chain. But not for long. As more fossils were uncovered, the chain splayed out into the usual phylogenetic net, and it was all too apparent that evolution had not been in a straight line at all, but that (to consider size only) horses had now grown taller, now shorter, with the passage of time. Unfortunately, before the picture was completely clear, an exhibit of horses as an example of orthogenesis had been set up at the American Museum of Natural History, photographed, and much reproduced in elementary textbooks (where it is still being reprinted today).[13]

Hardin is right in saying that this misleading picture is still being reprinted today, if only in elementary textbooks rather than in professional literature. I recently found a textbook published by Barnes & Noble in 1962 entitled *Fossils, An Introduction to Prehistoric Life* by William H. Matthews III, professor of geology at Lamar State College of Technology in Beaumont, Texas. There on page 43 was the famous picture of the horses, properly credited to the American Museum of Natural History. I wrote to Professor Matthews, calling his attention to Hardin's remarks. He replied: ". . . the audience for whom this book is intended has very little scientific background and it is for this reason that the material is handled as it was. In more technical discussions, of course, we consider the approach suggested by Professor Hardin."

I find this double standard improper and unwise, and hence am happy to add that it has not been adopted by the entire profession. In 1967 my daughter showed me the textbook used in her freshman geology course at Smith College.[14] The author, a Yale professor emeritus, printed a similar picture and credited it to the Yale Peabody Museum. I wrote to the Museum, quoting Hardin and asking how I could explain the situation to my daughter. I received a prompt and courteous reply from Professor Karl M. Waage, saying that the picture was perhaps misleading in an elementary book;* that he would be collaborating on the next edition; and that he would certainly make a change. His embarrassment was obvious.

The reader will now be able to understand why I want to make appropriate reservations about the implications of comparative anatomy and embryology. They are splendid studies, but they serve only to pose problems. They do not solve them. Solutions, if they exist, are in the province of evolutionary theory and will not be furnished by the narrower branches of study.

* A picture like this is useful as a comparison of anatomical forms, but is misleading because it is usually interpreted (and not only by laymen) as a line of descent. It is what is technically known as a *Stufenreihe* (series of stages), but is taken as an *Ahnenreihe* (series of ancestors). Hence Dr. Waage's use of *perhaps*.

1. Mayr (1963), 3.
2. Deevey (1967), 639.
3. Darwin (1859), 439–450.
4. Darwin (1859), 449.
5. Smith (1958), 265–266.
6. Bateson (1894), 8–10; Smith (1958), 264; Eiseley (1961), 95–97.
7. Quoted in Stebbins (1950), 488.
8. Simpson (1949), 218.
9. Stebbins (1950), 488.
10. Huxley (1957), 19.
11. Hardin (1961), 79–86.
12. Olson (1960), 536, 540.
13. Hardin (1961), 225–226.
14. Dunbar (1960), 412.

One who makes a close study of almost any branch of science soon discovers the great illusion of the monolith. When he stood outside as an uninformed layman, he got a vague impression of unanimity among the professionals. He tended to think of science as supporting an Establishment with fixed and approved views. All this dissolves as he works his way into the living concerns of practicing scientists. He finds lively personalities who indulge in disagreement, disorder, and disrespect. He must sort out conflicting opinions and make up his own mind as to what is correct and who is sound.

This applies not only to provinces as vast as biology and to large fields such as evolutionary theory, but even to small and familiar corners such as the species problem. The closer one looks, the more diversity one finds. There is tension between zoologists and botanists. Paleontologists and geneticists see things very differently. Religion and philosophy exert cryptic influences. One finds no trace of a terse and noncontroversial formulation of the species problem such as the editor of an encyclopedia must long for.

Let us begin with the men who work with species questions day in and day out. These are the taxonomists or systematists. They perform the valuable and necessary task of classifying plants and animals into a hierarchical system of groups. Species that are more like each other than like other species are grouped together in a genus; similarly, genera are grouped into families; , families into orders; orders into classes; and classes into phyla. A species is a small unit such as the marigold or the camel, whereas a phylum is a vast group such as the vertebrates.

This is not easy work. The taxonomist must have a perfect acquaintance with his own special field; he must be thoroughly grounded in many branches of biology, such as evolutionary theory, genetics, cytology, and ecology; and he must master the techniques of biometry and statistics. Above all, he must have unremitting patience and a strong back. The ideal taxonomist is a superman.[1]

Needless to say, not all the practicing taxonomists are ideal. Instead of being supermen, they suffer from very human failings. One of these is that, owing to temperament or training, they tend to be "splitters," who multiply species by seizing on small differences, or "lumpers," who bundle many different forms together under one name. Thus one man recognized about 200 species of snail in Hawaii in 1905, while a three-man team in 1912 reduced the total to 43.[2] Another failing is the tendency to refine the vocabulary in a bewildering way, subdividing the basic unit into varieties, races, subspecies, semispecies, sibling species, and so on, although there are no clear standards for delimiting these categories. A further cause of irritation to some, although it is more a professional necessity than a failing, is that the taxonomists tend to live comfortably in museums and laboratories (ivory towers to their critics) instead of doing field work in New Guinea or Patagonia.

The result is that there is no clear-cut and easily applied system of classification, and the taxonomists are reproached by their fellow biologists. Professor Deevey of Yale calls them "contentious pedants." [3] Professor Hardin of the University of California sneers at "stamp-collector taxonomists still cluttering up the literature with unenlightening discussions of 'the species problem' as it is dignified." [4] Professor Simpson of Harvard condemns them wholesale: "Primate classification has been the diversion of so many students unfamiliar with the classification of other animals that it is, frankly, a mess." [5]

This must be discouraging to the taxonomists, but there are a few biologists who go out of their way to say a kind word. Thus Smith remarks sympathetically: ". . . taxonomists . . . are faced by a contradiction between the practical necessity and the theoretical impossibility of their task. In struggling with this contradiction, they have been led to make important contributions to our knowledge of evolutionary processes." [6] And David Lack, when he returned from studying Darwin's finches in the Galápagos Islands, gratefully acknowledged that the museum taxonomists, who classified birds on the basis of their skins without ever seeing the living creatures, were almost always right.[7] These tributes must compensate for some of the vituperation.

The original and supreme taxonomist was Carolus Linnaeus (1707–1778) of Sweden. Using the species-phylum hierarchy, he and his followers had worked out exhaustive classifications of plants and animals long before Darwin was born. Their building block was the species, and they regarded the species as something fixed and immutable. Their system was challenged and tested when the evolutionists entered the scene and declared that all was in flux, that nothing was fixed and immutable, and that classification was a hopeless task. This view is neatly expressed by Smith:

The theory of evolution holds that existing plants and animals have originated by descent with modification from one or a few simple ancestral forms. If this is true, it follows that all the characteristics by which we can classify them into species have been and are changing, and further that on many occasions in the past a single population has given rise to two or more populations whose descendants today are sufficiently different from one another to be classified as different species. Now there is no reason to suppose that either the processes of modification in time, or the processes of division of a single species into two, have always, or even usually, occurred in a series of sharp discontinuous steps. Therefore any attempt to group all living things, past and present, into sharply defined groups, between which no intermediates exist, is foredoomed to failure.[8]

After some initial fluttering, the taxonomists responded to the challenge in a sensible way. Without trying to settle the merits of evolutionary theory, they pointed out that their work had to go on if the plants and animals were ever to be described in a manageable way, and that the charts of relationship were pretty much the same no matter what theory the classifier believed in.[9] They are still frequently urged to look at evolutionary relationships[10] (also known as phylogenies), but this is more easily said than done when the relationships are largely unknown;[11] hence as a practical matter the taxonomists continue to rely, as they have always done, almost entirely on structure.[12] This adds to the irritation of the evolutionists.

Let us now turn to the basic question—What is a species? The answer is neither clear nor easy. Hardin says bluntly: ". . . no thoroughly satisfactory definition of a species can be given."[13] G. Ledyard Stebbins, Jr., Hardin's botanical colleague at the University of California, after pointing to twelve different definitions in the recent literature, takes comfort from the fact that there is at least a "large common ground of agreement among them."[14] One reputable scientist attempted to solve the problem by asserting that a species was what a taxonomist was willing to classify as a species: " . . . a species is a community, or a number of related communities, whose distinctive morphological characters are, in the opinion of a competent systematist, sufficiently definite to entitle it, or them, to a specific name."[15] I assumed that this was a sort of joke until

I found Sir Julian Huxley calling it "a quite reasonable definition of the term species." [16]

The difficulties in defining species are troublesome enough, but they are overshadowed by the fact that there are two widely varying philosophical approaches to the species question. These approaches go back to Plato. One school of thought (seldom fully articulated)[17] regards a species as something real, as Linnaeus did; while another school regards the species name as only a convenient label. The two schools can be called the realists and the nominalists,[18] as was done in the disputes of the medieval scholastics. The realists are accused of sympathy for archetypes, types, and platonic ideas. The nominalists are accused of an atomism that undermines all systems and generalizations.

This difference of opinion is not solely due to the temperaments and dispositions of the taxonomists. Nature herself has set the stage by her whims and inconsistencies, especially by providing excellent examples for each line of thought. Those who contend that species have an objective reality select examples such as the ginkgo tree, which is a very ancient and regular form with few varieties and no close relatives. Those who argue that species do not exist in nature point to the willows, which have endless varieties and hybrids shading into each other.[19] The ginkgo is a loner; the willows are a continuum. No wonder that schemes of classification find it hard to accommodate such disparate material.

This cleavage produces some vigorous argumentation. Professor Mayr of Harvard is fiercely opposed to what he calls "typological thinking" and rejoices in "eliminating the last remnants of Platonism, by refusing to admit the *eidos* (idea; type, essence) in any guise whatsoever." [20] Eiseley shuns Plato as he would the plague.[21] On the other hand, the dangers of nominalism were pointed out as early as 1860, when Louis Agassiz, Mayr's eminent predecessor at Harvard, asked: "If species do not exist at all, as the supporters of the transmutation theory maintain, how can they vary? And if individuals alone exist, how can the differences which may be observed among them prove the variability of species?" [22] These are, as Mayr freely admits,[23] completely valid questions, since it is hard to see what we are talking about if there are no species. The nominalists find them hard to answer; in fact, it is only in recent years that any serious answer has been essayed.

The proposed answer is a sort of compromise. It gets away from Agassiz by defining a species as an interbreeding population and denying any superindividual units. It gets away from Plato by loudly disclaiming him and denying any inclination to look beyond the physical individuals in the population. It directs itself to *population* rather than *typological* thinking.[24] It is hailed by Mayr, Simpson, Dobzhansky, and other leading members of the synthetic school as one of their greatest achievements.[25]

Nevertheless, all is not well with the new solution. Aside from possible doubts as to its ability to withstand analysis by philosophers, even its proponents concede that it has three glaring weaknesses. First, when interbreeding is the decisive criterion, groups may be regarded as different species even when they look identical. This has actually happened with the famous fruit fly *Drosophila*, where four groups are counted as separate species although no one can tell them apart under the microscope.[26] Thus the scientists have abdicated and allowed the flies to classify themselves by their own whims. Second, nothing is known about interbreeding among the vast number of extinct plants and animals, and hence the paleontologists are not helped by the definition. Third, in large areas of the plant and animal kingdoms, there is no breeding at all in a sexual sense, progeny being provided by other methods such as agamospermy and vegetative reproduction: hence the criterion is again useless.[27] This seems to be why Stebbins, the botanist, cannot go along with the synthetics (most of whom are zoologists or geneticists) although he sees no satisfactory alternative.[28] This may also be why the International Code of Botanical Nomenclature, through at least four editions, studiously refrained from defining the word "species." [29]

It is perhaps surprising that a theory of species should be propounded in the face of such obvious deficiencies, but the literature is clear and the reason is not hard to find. Mayr, the leader in this field, recognizes that nature is so complex, diverse, and inconsistent that "no system of nomenclature and no hierarchy of systematic categories is able to represent adequately the complicated set of interrelationships and divergences found in nature." [30] In spite of this, however, as a practicing biologist Mayr demands a concise definition of the *species*, because he cannot work without it. If a man insists on defining what he knows to be undefinable, he can hardly avoid falling into the grotesque.

What was Charles Darwin's position on this question? There is good evidence that he was a nominalist, since he said: "I look at the term species, as one arbitrarily given for the sake of convenience to a set of individuals closely resembling each other." [31] Mayr,[32] Lack,[33] and many others take this at face value and assert that Darwin was a nominalist. On the other hand, Sir Ronald Fisher, the leader among the mathematical biologists, chides certain writers for saying that Darwin held that "the word *species* does not correspond to any existing reality." [34] He thinks Darwin supported inconstancy but not nonexistence, and hence was not a thorough nominalist.

This is not the whole story. In another passage, Mayr says that Darwin had an "essentially *typological* species definition." [35] This seems to move Darwin into the realist camp, although Mayr had previously ranked him as a champion of nominalism; hence I was inclined to regard it as a slip of the tongue, until I noticed some suggestive language in *The Ori-*

gin of Species. Darwin quotes, with obvious approval, a gentleman who, speaking of what the breeders had done for sheep, said: "It would seem as if they had chalked out upon a wall a form perfect in itself, and then had given it existence." Darwin adds admiringly: "In Saxony the importance of the principle of selection in regard to merino sheep is so fully recognized that men follow it as a trade: the sheep are placed on a table and are studied, like a picture by a connoisseur." [36] Lurking within these phrases there seems to be a leaning toward archetypes or ideal forms, which is exactly what Mayr repudiates as typological thinking.* Perhaps Darwin actually was a realist. Perhaps he fluctuated.[37] Perhaps he was not fully aware of the arguments. In any event, he never heard of the interbreeding-population theory.

I spoke earlier of lively personalities and disrespect. Let me furnish some examples. Mayr, a convinced evolutionist and an eminent member of the synthetic school, says that Darwin was "bewildered," [38] that he was "hopelessly confused," [39] and that he had a "lack of understanding of the nature of species." [40] He adds that Darwin was unable to discover the origin of species: "Darwin failed to solve the problem indicated by the title to his work. Although he demonstrated the modification of species in the time dimension, he never seriously attempted a rigorous analysis of the problem of the multiplication of species." [41] Professor Simpson, Mayr's colleague at Harvard and an equally convinced evolutionist, caps Mayr by saying that Darwin's "book called *The Origin of Species* is not really on that subject." [42] What would Darwin say to this?

The biologists have very difficult problems, which I am glad to be under no obligation to solve. I can understand why Darwin was hopelessly confused, though sometimes I suspect that

---

* It is hard to find a scientist who will defend the concept of archetypes, although many obviously have a leaning in that direction. The best presentation known to me is Charles Williams's novel *The Place of the Lion* (Pellegrini & Cudahy, 1951).

Mayr is equally confused. But one thing is clear—the experts do not pretend that classical Darwinism was either clear or correct as to the species problem. That is sufficient for the thesis of this book.

1. Stebbins (1950), 4–7; Huxley (1942), 156–159, 390–402.
2. Robson (1928), 89.
3. Deevey (1967), 639.
4. Hardin (1961), 76.
5. Simpson (1949), 81.
6. Smith (1958), 153–154.
7. Lack (1947), 16.
8. Smith (1958), 152.
9. Dupree (1959), 386; Smith (1958), 31.
10. Stebbins (1950), 6–7.
11. Huxley (1942), 374–375, 400; Rickett (1968), 13; Darwin (1871), 514.
12. Abercrombie, Hickman, and Johnson (1966), 61.
13. Hardin (1961), 76.
14. Stebbins (1950), 189.
15. Dr. C. Tate Regan, director of the Natural History Museum at South Kensington; quoted in Huxley (1942), 157.
16. Huxley (1942), 157; Mayr (1942), 115, also takes it seriously.
17. Huxley (1942), 167, quotes several writers who have a vague feeling of reality. Williams (1966), 253–254, agrees.
18. Simpson (1953), 340; Grant (1957), 44.
19. Grant (1957), 65; see also Simpson (1950), 257–261.
20. Mayr (1966), xi; (1963), Chapters 2 and 3; (1959) throughout.
21. Eiseley (1960), 23, 48, 65.
22. Agassiz (1860), 143.
23. Mayr (1959), 195. Ghiselin (1969), 92, indicates that Darwin agreed with Agassiz.
24. Mayr (1966), xix–xx.
25. Dobzhansky (1956), 337–347; Mayr (1963), 12–30; Simpson (1964), 72.
26. Dobzhansky (1956), 337; Sonneborn (1957), 195–196.
27. Stebbins (1950), 383; Grant (1957), 50–51.
28. Stebbins (1950), 201–203; Huxley (1942), 396.
29. Grant (1957), 42–43.

30. Mayr (1942), 103; see also 113–114, 147, 151, 172, 190, and 200. For the demand for a definition, see 113–114.
31. Darwin (1859), 52.
32. Mayr (1950A), 175, and (1963), 14.
33. Lack (1947), 125.
34. Fisher (1954), 87–88.
35. Mayr (1963), 484.
36. Darwin (1859), 31.
37. Ghiselin (1969) shows a series of fluctuations at 89, 91, 92, 93, 94, 101, 102, and 149.
38. Mayr (1963), 13.
39. Mayr (1963), 484.
40. Mayr (1963), 12.
41. Mayr (1963), 12.
42. Simpson (1964), 81; see also Huxley (1942), 153, 387.

The experience of breeders was of deep interest to Darwin. He bred pigeons himself and hobnobbed with pigeon fanciers. He spent a great deal of time talking to breeders of all sorts and recording their observations in his copious notebooks. He was familiar with the great improvements that had been made in many plants and domestic animals. Change was occurring before his eyes. What could be more encouraging to a man who was brooding on the idea of evolution?

But there was a difficulty. The observed changes were small. The breeders could improve a sheep's wool or create a larger rose, but they never even tried to make big changes, such as adding wings to a horse.

I am going to use the terms "micro" and "macro" to describe small changes and large. Most small changes concern varieties, such as toy poodles or giant aspidistras, but those who take a narrow view of species may say that such changes affect species or even genera. There is no exact line between these classifications, and there is no exact line between small and large variations; but any sensible person can see that there is a difference between small and large, especially if we assist him with crass examples such as the contrast between breeding black horses and breeding winged horses.

The changes that Darwin observed in the breeding pens were all micro. They occurred without question, but they were not sufficient for his purposes when he was faced with macro gaps between his units (the types or species), because all of these started out with distinct forms even in the earliest fossils. Comparative anatomy and embryology showed resemblances between the units, but they also showed that between the units there were gulfs going back to the misty beginnings. Looking only at large domestic quadrupeds, it was easy to see that horses, cows, sheep, and goats all had a backbone, four limbs, a brain, a heart, a skull, and a reproductive system, and that these members were similar in many ways; but no one would say that these animals were identical. They looked like cousins, but there was neither a neatly graduated series of living links between them nor a converging fossil genealogy behind them. Darwin had to find processes by which the gaps could be bridged.

DARWIN entertained the very questionable opinion that animals and plants could vary in all directions and to an unlimited degree. In the first edition of *The Origin of Species* he said: "I can see no difficulty in a race of bears being rendered, by natural selection, more and more aquatic in their habits, with larger and larger mouths, till a creature was produced as monstrous as a whale." [1] He knew that this was not the common view, since as early as 1844 he had written: "That a limit to variation does exist in nature is assumed by most authors, though I am unable to discover a single fact on which this belief is grounded." [2] He neglected to add that he also could not discover a single fact on which an opposite belief might be grounded.*

* One author whom Darwin must have had in mind was T. R. Malthus, whose *Essay on the Principle of Population* influenced Darwin profoundly when he first read it in 1838. In Chapter 1 of Book 3 of this work, Malthus took issue with those who contended that they could improve plants and animals as much as they liked. He pointed out that a variety of sheep had been bred for small head and legs, but that it could hardly be carried to a

Darwin was a timid man in many ways, but fortified by his faith in variation he acted boldly in this situation. He took the micro changes observed by the breeders (which in themselves did not begin to fill the gaps) and he *extrapolated* them. He said, in brief, that twenty years of breeding often achieved substantial changes; therefore, if nature continued the work for a hundred million years, it could close all the gaps. His actual phrasing was more poetic: "Slow though the process of selection may be, if feeble man can do so much by his powers of artificial selection, I can see no limit to the amount of change, to the beauty and infinite complexity of the co-adaptations between all organic beings, one with another and with their physical conditions of life, which may be effected in the long course of time by nature's power of selection." [3]

Extrapolation is a dangerous procedure.[4] If you have a broad base of sound observations, you can extend it a little at the ends without too much risk; but if the base is short or insecure, extension can lead to grotesque errors. Thus if you observe the growth of a baby during its first months, extrapolation into the future will show that the child will be eight feet tall when six years old. Therefore all statisticians recommend caution in extrapolating. Darwin, however, plunged in with no caution at all.

Despite Darwin's easy confidence, it seems likely that his extrapolation was not justified. The first difficulty is that no one has ever seen a macro change take place, whether in the breeding pens or among the fossils. My paperbacks seemed to concede this, but the point worried me so much that I spent ten dollars for *Evolution Above the Species Level* by the German biologist Bernhard Rensch. I found that Professor

---

point where the head and legs disappeared entirely or were reduced to the scale of a rat. He added that a carnation would never produce a flower as big as a large cabbage. These statements are negatives that cannot be proved, but they are so reasonable that surely Darwin has the burden of proof when he takes the opposite position.

Rensch did not pretend to have any actual examples in hand, although he asserted that macro changes (which he prefers to call transspecific evolution) should not be regarded as impossible.[*]

The next difficulty is the lack of transitions. If we join Darwin in assuming that macro changes *must* have been accomplished by small steps, so that the gaps were at one time filled, then what has happened to all the intermediate forms? This question occurred to Darwin, and he furnished the answers that are still in use today—the extreme imperfection of the geological record and the poorness of our paleontological collections.[5] Hardin, asking himself a hundred years later whether he can show all the links in the chain, replies: "No, of course not; the geological record is imperfect and will always remain so, since it is highly improbable that short-lived intermediate species will be fossilized." [6]

This is the standard answer, but it is rather threadbare after a century of digging and collecting. The simple phrase "short-lived" is already troublesome. How does Hardin know they were short-lived if he has never seen them? Can any species really be short-lived when Huxley, reflecting the generally accepted view, says that large changes occur over tens of millions of years, while really major ones (what we would call macro) take a hundred million or so? [7]

The heart of the problem is whether living things do indeed vary to an unlimited extent or, to state it differently, whether micro changes cumulate into macro effects. The instinctive feeling of untutored men is against this. The species look stable. We have all heard of disappointed breeders who carried their work to a certain point only to see the animals or plants

---

[*] Professor Rensch's effort to demonstrate nonimpossibility is an illustration of what Fischer (1970), 53, has in mind when he says: "*The fallacy of the possible proof* consists in an attempt to demonstrate that a factual statement is true or false by establishing the possibility of its truth or falsity. This tactic . . . never proves a point at issue. Valid empirical proof requires not merely the establishment of possibility, but an estimate of probability."

revert to where they had started. Despite strenuous efforts for two or three centuries, it has never been possible to produce a blue rose or a black tulip.[8] Darwin himself knew in 1844 that most authors assumed there were limits to variation, and he also knew that among pigeons the crossing of highly bred varieties was apt to provoke a reversion to "the ancient rock-pigeon." Was he discouraged when, in the sixth and last edition of *The Origin of Species,* he quietly excised the above passage about converting bears into whales?

But it is not only untutored men and pre-Darwinian authors who are skeptical. Eiseley reports the discovery by the Danish scientist W. L. Johannsen that "the variations upon which Darwin and Wallace had placed their emphasis cannot be selectively pushed beyond a certain point, that such variability does not contain the secret of 'indefinite departure.' " [9] Sir Julian Huxley reports that, in a pure eyeless strain of fruit flies, after eight or ten generations the eyes had reverted almost to normal.[10]

I was also impressed by the story of a biologist who broke with orthodox theory. The late Richard B. Goldschmidt (1878–1958) must have been a highly tutored man, since Hardin calls him an "important geneticist" [11] and Smith devotes several pages to him.[12] After observing mutations in fruit flies for many years, Goldschmidt fell into despair. The changes, he lamented, were so hopelessly micro that if a thousand mutations were combined in one specimen, there would still be no new species.[13] This led him to propose the hypothesis of the "hopeful monster," whereby a huge change might have occurred all at once and been preserved by a favoring environment. His colleagues rejected this proposal as unsound, but they seem to escape Goldschmidt's despair only by an act of faith.*

* There are moments when Simpson seems to be in basic agreement with Goldschmidt, although he speaks of "quantum" evolution rather than

These pieces of evidence led me to suspect, diffidently at first, that extrapolation was up to its old tricks, that micro changes did not aggregate into macro, that macro changes could not be shown to occur, and that one of Darwin's main props had collapsed. While I was wrestling with this suspicion I encountered the works of Ernst Mayr of Harvard, who has become one of my principal sources.

Mayr notes that animal populations have a certain persistence or inertia, in that they resist sudden or drastic change, and he gives this persistence the elegant name of "genetic homeostasis." He also provides a splendid example of what I had been groping for—the corollary tendency of animals and plants to balk at being bred too far in any direction. This comes out in his description of some work in 1948 with the famous fruit fly, *Drosophila melanogaster*.[14] Here is the gist of his account.

Two experiments were run, one for decrease and one for increase in the number of bristles, which averaged 36 in the starting stock. Selection for decrease was able, after thirty generations, to lower this average to 25 bristles, but then the line became sterile and died out. A mass low line (maintained without selection) was started with 32 bristles and remained nearly stable for ninety-five generations. All attempts to derive from this line others with lower bristle numbers failed because the lines died out before selection had made much progress. In the high line, progress was at first rapid and steady. In twenty generations the average rose from 36 to 56. Then sterility became severe and a mass line (without selection) was started. Average bristle number fell sharply and was down to 39 in five generations.

Mayr regards these results as entirely normal. He believes that there is just so much variability in a fruit fly, and that if it is pushed hard in one direction it will be distorted in an-

"macro." But the whole problem is left fallow by Simpson and largely ignored by his colleagues. This story is expanded in Chapter 17.

other. His language is plain: "Obviously any drastic improvement under selection must seriously deplete the store of genetic variability. . . . The most frequent correlated response of one-sided selection is a drop in general fitness. This plagues virtually every breeding experiment." [15]

Genetic homeostasis makes even micro changes look difficult, and seems to be a fatal obstacle to macroevolution. Nevertheless, Mayr himself continues to believe that macroevolution *must* take place through natural selection working on small changes. But he cites no observed cases; he confesses that he is relying on extrapolation; and in the midst of his tentative suggestions about a modus operandi, he concedes that "much of this is obviously speculative." [16] Thus he seems to be a reluctant but impressive witness against the cumulation of micro changes.

Mayr, with a century of literature at his fingertips, is immensely sophisticated. But Darwin, in the first dawn, already perceived the phenomena that Mayr describes. As to the idea of a limited store of variability, he quotes Goethe's perspicuous remark that "in order to spend on one side, nature is forced to economize on the other side." [17] As to the dangers of sterility, he said: "Sterility has been said to be the bane of horticulture." [18]

Having slowly concluded that there is no evidence that micro changes cumulate into macro effects, I was relieved to find that, although the subject is seldom discussed, my view is shared by reputable scientists.* Thus Eiseley says: "It would appear that careful domestic breeding, whatever it may do to improve the quality of race horses or cabbages, is not actually in itself the road to the endless biological deviation which is evolution. There is great irony in this situation, for

---

* Some years after reaching this conclusion, I was further relieved by my discovery of the Broom-Huxley doctrine that evolution is now exhausted (see Chapter 15). These scholars seem to assert flatly and confidently exactly what I had laboriously worked out for myself—that we see only micro evolution and that the micro steps do not cumulate into macro effects.

more than almost any other single factor, domestic breeding has been used as an argument for the reality of evolution." [19] Professor Deevey supplies terse phrases such as "the species barrier" and "the limited charter" to describe the situation, then confesses bankruptcy: "Some remarkable things have been done by crossbreeding and selection inside the species barrier, or within a larger circle of closely related species, such as the wheats. But wheat is still wheat, and not, for instance, grapefruit; and we can no more grow wings on pigs than hens can make cylindrical eggs." [20] Thus my surmise about winged horses is confirmed in New Haven.

When the experience of breeders is in question, it is prudent to consult competent breeders. Luther Burbank who, though no theoretician, was the most competent breeder of all time, looked at this problem. He eloquently endorsed the limited charter:

There is a law . . . of the Reversion to the Average. I know from my experience that I can develop a plum half an inch long or one 2½ inches long, with every possible length in between, but I am willing to admit that it is hopeless to try to get a plum the size of a small pea, or one as big as a grapefruit. I have daisies on my farms little larger than my fingernail and some that measure six inches across, but I have none as big as a sunflower, and never expect to have. I have roses that bloom pretty steadily for six months in the year, but I have none that will bloom twelve, and I will not have. In short, there are limits to the development possible, and these limits follow a law. But what law, and why?

It is the law that I have referred to above. Experiments carried on extensively have given us scientific proof of what we had already guessed by observation; namely, that plants and animals all tend to revert, in successive generations, toward a given mean or average. Men grow to be seven feet tall, and over, but never to ten; there are dwarfs not higher than 24 inches, but none that you can carry in your hand. . . . In short, there is undoubtedly a pull toward the mean which keeps all living things within some more or less fixed limitations.[21]

The dangers of extrapolation became very evident to Simpson when he tried to calculate the tempo of evolution. Working from what he knew of the fossils and time sequences, he could see that the bat's wing, for instance, had changed very little since the middle Eocene (about one hundred million years ago). If its earlier evolution had proceeded at the same slow rate, its total time of development would be greater than the age of the earth, a manifest absurdity. Therefore Simpson concluded that in the early days the rate for bats must have been ten to fifteen times as fast as later.*

Despite this testimony showing the species barrier and the dangers of extrapolation, some biologists continue to extrapolate as ardently as ever Darwin did. Thus Sir Julian Huxley says: "With the length of time available, little adjustments can easily be made to add up to miraculous adaptations; and the slight shifts of gene frequency between one generation and the next can be multiplied to produce radical improvements and totally new kinds of creatures." [23] I found that Professor John Tyler Bonner of Princeton was equally bold: "There is no reason to believe that these large changes are not the result of the very same mechanisms as the small changes. . . . One involves a small step over a few years; the other involves many many thousands of steps over millions of years." [24] Huxley and Bonner do not seem to be familiar with genetic homeostasis, although Huxley knows of the disappointments with blue roses and black tulips.

Having quoted Luther Burbank, I will now depart even further from professional scholarship by quoting Mark Twain's views on extrapolation:

In the space of 176 years the Lower Mississippi has shortened

---

* The fragile nature of these speculations should be carefully noted. The actual fossil record shows very slow change and leaves little time available. Therefore Simpson is forced to assert (without evidence) a rapid rate before the curtain went up. But if he made it *too* rapid, he would be approaching the sudden leap, or saltation, to which he is unalterably opposed. [22]

itself 242 miles. That is an average of a trifle over a mile and a third per year. Therefore any calm person who is not blind or idiotic can see that in the Old Oölitic Silurian Period, just a million years ago next November, the Lower Mississippi River was upward of 1,300,000 miles long and stuck out over the Gulf of Mexico like a fishing-rod. And by the same token any person can see that 742 years from now the Lower Mississippi will be only a mile and three-quarters long, and Cairo and New Orleans will have joined their streets together and be plodding along comfortably under a single mayor and a mutual board of aldermen. There is something fascinating about science. One gets such wholesale returns of conjecture out of such a trifling investment of fact.

I doubt that Huxley and Bonner have heard of Mark Twain's calculations, but Deevey is familiar with them and with the dangers of extrapolation. Therefore it is astonishing to me that he should brush them off by saying: "Yet a yachtsman makes extrapolations just as breathtaking whenever he consults his watch and waits for high tide before sailing." [25] The tide tables are based on a hundred years of daily observations, which need be projected not more than twelve and one-half hours into the future to give the next high tide. The extrapolations of Huxley, Bonner, and Darwin are the other way round; they are based on a few years of observation and are projected hundreds of millions of years into the past and future. The two cases are not equally breathtaking.

I cannot assert that the biologists have expressly abandoned Darwin's position. Indeed, it seems likely that most of them would say that he simply *must* be correct. But on the other hand, they would all recognize the limits of variability, the curse of sterility, the dangers of extrapolation, the hopelessness of trying to convert bears into whales or of breeding winged horses, and the strong inertia of genetic homeostasis. I do not see how these points can be reconciled with Darwin's position, and I suggest that the time has come for a retreat.

1. Darwin (1859), 184.
2. Eiseley (1958), 186.
3. Darwin (1859), 109.
4. Olson (1960), 532–534, is the fullest and most temperate discussion I have found of the dangers of extrapolation. He urgently recommends caution. Simpson (1964), 140–141, also warns against extrapolation, though Simpson frequently extrapolates. See also Fischer (1970), 120–122.
5. Darwin (1859), 280, 287.
6. Hardin (1961), 103.
7. Huxley (1957), 13.
8. Huxley (1942), 519.
9. Eiseley (1958), 227.
10. Huxley (1953), 40.
11. Hardin (1961), 226.
12. Smith (1958), 276–284.
13. Goldschmidt (1952), 94.
14. Mayr (1963), 285–286.
15. Mayr (1963), 290.
16. Mayr (1963), 586, 613, 615.
17. Darwin (1859), 147.
18. Darwin (1859), 9.
19. Eiseley (1958), 223.
20. Deevey (1967), 636.
21. Quoted in Hall (1939). Another great breeder demonstrated that the tendency to vary does not operate in all directions and to an unlimited degree; see N. I. Vavilov (1951) and his study of homologous variation; also Huxley (1942), 511, 519. Compare Francis Galton's law of filial regression; Kellogg (1925), 122.
22. Simpson (1944), 119, 139; (1953), 351–353.
23. Huxley (1957), 41.
24. Bonner (1962), 48.
25. Deevey (1967), 637.

Darwin never tried to define natural selection in a rigid way, but it is fairly clear that for him it was not a complex concept. It amounted to little more than the fact that, for various reasons, among all the individuals produced in nature some die soon and some die late. Thus natural selection, for Darwin, was *differential mortality*.[1] In the course of time there has been a slow change in this view, so that now it is customary to say that natural selection is *differential reproduction*.[2] This in turn may be equated with reproductive success, or leaving the most offspring.

The difference between these formulations is neatly illustrated by Simpson: "Suppose that all the individuals in a population lived for precisely the same length of time, with no elimination of the unfit or survival of the fittest, hence no Darwinian selection. Suppose further that . . . the taller ones, or those with an allele A, or a chromosome arrangement M, or a hereditary fondness for apples, had twice as many offspring as those without these characteristics. Then there would be very strong, clearly non-Darwinian selection."[3]

Is natural selection the sole factor in evolution? Sir Julian Huxley says yes: "So far as we now know, not only is Natural Selection inevitable, not only is it *an* effective agency of evolution, but it is *the* only effective agency of evolution." [4] Almost every other author, however, has his own list of further factors. Thus Rensch of Münster speaks of mutations, recombination of genes and gene flow, fluctuations of population, and processes of isolation, as well as processes of selection.[5] Simpson mentions variability, rate and character of mutations, length of generations, and size of populations, as well as natural selection.[6] Stebbins, the botanist from the University of California, names fluctuations in population, random fixation of genes, isolation, and natural selection.[7] More lists could be found, but natural selection would be prominent in all of them. It is the commonest and most potent phrase in the Darwinian vocabulary.

I was startled to find that, although natural selection is included in all lists, there has been wide dispute as to its importance. The early Darwinians thought that every aspect of every animal, right down to the number of spots or bristles, was determined by natural selection and was therefore "adaptive," i.e., important for survival. By rashly undertaking to explain just why these trivial features were adaptive, the enthusiasts got themselves entangled in wild speculations and absurd reasoning. Sir Julian Huxley sums up the failings of the energetic explicators: "The paper demonstration that such and such a character was or might be adaptive was regarded by many writers as sufficient proof that it must owe its origin to Natural Selection. . . . There was little contact of evolutionary speculation with the concrete facts of cytology and heredity, or with actual experimentation." [8] Stebbins asserts that the whole idea fell into contempt: "In the early part of the present century . . . the prestige of the selection theory declined until many biologists regarded it not only as a relatively unimportant factor in evolution, but in addition as a subject not worthy of study by progressive, serious-minded biologists." [9]

The prestige of natural selection has risen greatly, but there is still a wide variety of opinion. Simpson says that some students ascribe almost no importance to it, while others believe it is the only really essential factor in evolution.[10] Stebbins laments that, because the "adaptive" nature of certain traits cannot be easily seen or proved, a number of reputable biologists argue that it does not exist, thus virtually denying natural selection.[11]

The reason for this diversity soon became clear to me. We are dealing with something invisible. The operations of natural selection, real or imagined, are not accessible to the human eye.

This first dawned on me when I found Stebbins saying (107): ". . . while the demonstration that selection has occurred is not excessively difficult, the nature of action and the causes of this selective process are much harder to discover or to prove." Being thus alerted to the presence of a problem, I watched closely as Stebbins circled around it. He recurred to it soon (118), first stipulating that we are obligated to determine what is adaptive, then confessing that this is impossible in the present state of the art: "Obviously . . . a final estimate of the importance of selection in evolution must depend largely on determining what . . . differences are . . . adaptive. . . . Unfortunately, however, the determination of the adaptive character of many types of differences between organisms is one of the most difficult problems in biology." The theme appears again on the next page: "These . . . changes in the composition of populations can prove convincingly the existence of natural selection as an active force, but the demonstration of how selection acts, and of the reason for the selective value of a particular character, is a much more difficult task." Stebbins comes back to the problem near the end of his book (506), but only to confess despondently: "We can, therefore, do little more than speculate."

Stebbins may be more troubled than most of his colleagues,

but they also agree that the problem is beyond them. Mayr says: ". . . one can never assert with confidence that a given structure does not have selective significance." [12] Dobzhansky, the great geneticist from Columbia, is equally firm: ". . . the value of anthropomorphic judgments on what constitutes a malformation is spurious." [13] Simpson even regards this as a matter of common knowledge: "The fallibility of personal judgments as to the adaptive value of particular characters, most especially when these occur in animals quite unlike any now living, is notorious." [14]

The mathematicians are partly responsible for this strange situation. They have demonstrated in an abstract way that infinitely small causes may have enormous effects in evolution, although it is impossible to observe the processes either in nature or in the laboratory. Simpson makes this clear in discussing a hypothetical case where animals with trait A survive one time oftener in ten thousand cases than animals with trait B: "By present techniques, it would be quite impossible to observe such weak selection either in the laboratory or in nature. . . . selection may be highly effective although quite beyond our powers of observation. . . ." [15]

If it is impossible to observe the processes either in nature or in the laboratory, no one can prove the mathematicians to be wrong. By the same token, of course, no one can empirically prove them to be right. Naturally, this dilemma is painful, and Simpson candidly concedes that ". . . it might be argued that the theory is quite unsubstantiated and has status only as a speculation." [16] From the evidence, I would have thought this had already been demonstrated rather than merely being arguable, but Simpson pursues this line of thought no further.*

---

* Some writers will allow only good Darwinians to slide away from this dilemma. When certain Lamarckians contend that inherited effects are so slight that they cannot be detected experimentally, Sir Julian Huxley

Even if we assume that natural selection exists and is continuously at work, it is impossible to determine the *intensity* of its action. It is not uncommon for biologists to say that in certain circumstances natural selection relaxes its vigilance and that in others it operates with unusual rigor,[17] but since there is no tangible evidence to justify such statements they seem to be only another way of saying that change is slow or fast. Simpson, although himself a frequent offender in this regard, admits: "The determination of intensity of selection is in itself a problem to which there is apparently no direct approach and one which it is very difficult to treat practically." [18]

Natural selection is almost always handled in general terms. Indeed, how could it be otherwise when its operations and intensity are beyond our ken? This means that it has no explanatory power when specific problems arise. I had sensed this in a vague way, but never saw it clearly formulated until I read Deevey's work. After describing a number of remarkable phenomena such as ultrasonar in bats and explosive charges in bombardier beetles, he says: "Of course these things are marvels, and of course, the fossil record being what it is, no one can say with confidence exactly how any one of them came about." [19] Note the word *exactly*. The Darwinians contend that any given result must have been produced by natural selection working on small changes, but when asked to be exact they are helpless.[20] Thus Dobzhansky cannot explain why the more than six hundred known species of *Drosophila* all have three orbital bristles on either side of their heads.[21] Sir Gavin de Beer admits that ". . . the causes of the origins of patterns, colors, and of many other things, are not known." [22] Simpson cannot explain why average stature in the United States has increased since 1900.[23] Simpson even confesses to ". . . the sad, one might almost say the shame-

---

(1942), 459, rebukes them roundly: "To plead the impossibility of detection is a counsel of despair. It is also unscientific."

ful, fact" that he does not know what natural selection is doing.[24]

The temptation to explain being what it is, I admired Deevey's willingness to view things as marvels and refrain from utilitarian explanations. I applauded Robert Ardrey when he voiced the same sentiment even more eloquently: "It is fruitless to attempt to explain everything in the natural world in terms of selective value and survival necessity. There are times when one can only record what is true, and dissolve in wonder." [25] I was disappointed to find Ardrey, a few pages later, succumbing to the temptation to explain, by the easy device of recognition marks, the strange phenomenon known as the prairie-dog kiss: "The kiss came about, I should assume, as a means of identification in the dark recesses of one's burrow to make sure by proper flavor that no stranger has sneaked in. Whatever its origin or selective value may be, whenever the members of a coterie meet, they exchange what is very nearly a human kiss, open-mouthed, and they seem to enjoy it." [26]

Simpson sets a much better example. When he sees one type of squirrel with ear tufts and another with none, he wonders but does not explain; and above all he does not comfort himself with the recognition-mark stratagem: ". . . why should *S. aberti* have handsome ear tufts that are quite lacking in *S. fremonti?* . . . It is always possible, and entirely true, to say that we just do not know the adaptive value of the differences, or we can guess at possible adaptive values, for which there is no evidence whatever, for instance that the ear tufts are 'recognition marks.' " [27]

The mention of Robert Ardrey reminds me of another troublesome aspect of our topic. Natural selection is supposed to be an impersonal force that replaces all Watchmakers or other guiding powers so that evolution can be explained without calling in any external agency.[28] Simpson allows himself to speak of the *opportunism* of evolution, but he is careful to warn us that "when a word such as opportunism is used, the reader should not read into it any personal meaning or anthropomorphic implications."[29] Here he puts his finger squarely on what worries me—the tendency of his colleagues to speak of natural selection in a personal or anthropomorphic way.

Robert Ardrey is the worst offender. He says at various points that natural selection is openminded; that it is not dogmatic; that it is blind as a cave fish, yet shrewd as a cat; that it has lost interest in the tooth; and that it regrets nothing.[30] I realize, of course, that Robert Ardrey is primarily a dramatist and that his sins must not be charged to the account of the professional biologists, yet he is only a little bit gaudier than many professionals. Darwin himself said: ". . . natural selection is daily and hourly scrutinizing . . . every variation, even the slightest; rejecting that which is bad, preserving and adding up all that is good; silently and insensibly working . . . at the improvement of each organic being. . . ."[31] Even Stebbins, the sober botanist, repeatedly speaks of natural selection as a *guiding* force or a *directive* force, and at one point he likens it to a sculptor creating a statue by removing chips from a block of marble.[32]

There may be no harm in using colorful language about an impersonal force, but it makes me uneasy. The Darwinians say they have banished the Watchmaker, but they may be raising his specter through their rhetoric.

My studies of natural selection had begun with no forebodings, but by this time I was becoming puzzled and skeptical. A process that operates invisibly, with an intensity that cannot be observed and with no ability to explain specific problems, an impersonal process that is continually given personal qualities—this sets my teeth on edge. Therefore I went back to the definitions to see if the premises were in order. I slowly realized that they were not. The phrase *differential reproduction* conceals a flaw.

The large and easy aspect of evolution is that some species have multiplied while others have remained stable and still others have dwindled or died out. As stated earlier, this is now conceded by everyone and needs no further demonstration. The problem is to explain why and how this occurs. About whether it does there has been no argument for many years.

If we say that evolution is accomplished largely by natural selection and that natural selection consists of differential reproduction, what have we done? Differential reproduction means that some species multiply by leaving more offspring than one-for-one, while others leave one-for-one and remain stable, and others leave less than one-for-one and dwindle or die out. Thus we have as Question: Why do some multiply, while others remain stable, dwindle, or die out? To which is offered as Answer: Because some multiply, while others remain stable, dwindle, or die out. The two sides of the equation are the same. We have a tautology. The definition is meaningless.

I regard this as a major discovery, a sort of lethal gene in the body of the central Darwinian doctrine; but I am not the first discoverer. It was formulated at least as early as 1959 by Professor C. H. Waddington of Edinburgh, a reputable member of the synthetic school, although his discovery seems to have had no impact on the biological fraternity. Waddington's statement is so staggering that it must be set forth in full:

> Darwin's major contribution was, of course, the suggestion that evolution can be explained by the natural selection of random variations. Natural selection, which was at first considered as though it were a hypothesis that was in need of experimental or observational confirmation, turns out on closer inspection to be a tautology, a statement of an inevitable although previously unrecognized relation. It states that the fittest individuals in a population (defined as those which leave most offspring) will leave most offspring. Once the statement is made, its truth is apparent. This fact in no way reduces the magnitude of Darwin's achievement; only after it was clearly formulated, could

biologists realize the enormous power of the principle as a weapon of explanation.[33]

Why do I find this staggering? Because a man who is astute enough to see that differential reproduction is a tautology is unable to see anything improper in a tautology. Because a man who reveres Darwin reduces Darwin's major contribution to a tautology, yet asserts that this does not reduce the magnitude of Darwin's achievement. Because a man who must know how weak natural selection is in explaining hard cases, and who has his finger on the reason for this weakness (the tautology), still speaks of the enormous power of natural selection as a "weapon of explanation."

Being now on the scent like a bloodhound, I studied the definitions further and soon hit pay dirt at Harvard. In his first two major works on evolution,[34] Simpson joined his colleagues in defining natural selection as differential reproduction, but there was a change in his third book. Perhaps dismayed by the difficulty of sorting out the active element in each case,[35] Simpson moved to a definition that rose above all difficulties, distinctions, and quibbles. He said: "I propose slightly to extend the definition used in population genetics and to define selection, a technical term in evolutionary studies, as *anything tending to produce systematic, heritable change in populations between one generation and the next.*" [36]

Where are the lists of influential factors now? What has happened to mutations, recombination of genes, processes of isolation, and length of generations? They are all gone, swallowed up by natural selection. This is the be-all and end-all. It is anything tending to produce change.*

---

* In the above definition Simpson emphasizes change. We get a glimpse of his versatility and of the slipperiness of the subject if we compare this with the definition at Simpson (1969), 127: ". . . natural selection . . . is . . . usually and most strongly a stabilizing, normalizing influence preventing or slowing and not hastening evolutionary change." The same view is expressed by Williams (1966), 54: "I regard it as unfortunate that the

But is such a broad definition of any use? We are trying to explain what produces change. Simpson's explanation is natural selection, which he defines as what produces change. Both sides of the equation are again the same; again we have a tautology.[37]

In discussing these questions with friends, I have found several who did not at once see anything wrong with a tautology. Therefore I will show the absurdities to which this one leads. Simpson studies rates of evolution and finds that they cannot be explained by mutations alone. Therefore, he says, an additional factor is necessary ". . . and the most reasonable probability is that that factor is selection." [38] But if selection is anything tending to produce change, he is merely saying that change is caused by what causes change. Thus clarified, even my friends can see that the net explanation is nil.

Again, on the next page, Simpson says that "ultimately it is the changes in environment that control rate of evolution, although the control is by means of the mechanism of selection." But how can selection have a mechanism when it is defined as anything tending to produce change? It has been diluted so far that no mechanism is left.

I then went back to Stebbins and his difficulty in discovering how selection acts and the reason for the selective value of a particular character. I perceived that his admissions were fatal. The argument nowadays is over the how and why of evolution, the question of whether having long since been decided in evolution's favor. As the how and why, Darwin offered natural selection. But if we now ask for the how and why of natural selection, Stebbins tells us that he cannot furnish them. He does not know the methods or the causes.

---

theory of natural selection was first developed as an explanation for evolutionary change. It is much more important as an explanation for the maintenance of adaptation." And again at 139: "So evolution takes place, not so much because of natural selection, but to a large degree in spite of it." Truly, natural selection is the be-all and end-all.

Therefore we must ask whether, aside from replacing evolution with natural selection as a term in the question, he has accomplished anything. The answer seems to be no.

If the reader is surprised to find natural selection disintegrating under scrutiny, I was no less so. But when we reflect upon the matter, is it so surprising? The biologists have innocently confessed that natural selection is a metaphor,[39] and every experienced person knows that it is dangerous to work with metaphors.[40] As the road to hell is paved with good intentions, so the road to confusion is paved with good metaphors. Perhaps the sober investigators should not have staked so much on a poetic device.

I cannot leave this subject without making a friendly and constructive suggestion. No one has asked me to do this, of course, but the views of a detached outsider can sometimes be of value. I am convinced that, although the biologists may not be able to see it at once, they have been led astray by an unjustified respect for the mathematicians.

I have no training in mathematics and cannot pretend to follow the operations of the mathematical biologists. But I suspect that the same is true of most of the run-of-the-mill biologists. Consider for instance a fairly typical passage from Sir Ronald Fisher's major work:[41]

> Often more than two genes may alternatively occupy the same locus. These are termed multiple allelomorphs. In extending the notion of genetic excess to such cases, it is convenient to define the genetic excess associated with a single gene. Thus if we suppose that the genotypic value X has been ascertained for an entire natural population, the genetic composition of each individual of which is known, we may let $\overline{X}$ stand for the general mean, and x for the deviation of any genotypic value, so that
>
> $$x = X - \overline{X}.$$
>
> Choosing any particular factor, we may pick out all the individuals carrying any one gene, counting the homozygotes twice,

and find the average value of x for this selected group of individuals.

Thus if out of a population of N individuals there are n_{11} homozygotes, and n_{lk} heterozygotes formed by combination with any other chosen allelomorph, the total of the values of x from the homozygotes may be represented by $S(x_{11})$, and that from any class of heterozygotes containing the chosen gene by $S(x_{lk})$. Then

$$\frac{2S(n_{11}) + \sum_{k=2}^{s}{}' S(n_{lk})}{2n_{11} + \sum_{k=2}^{s}{}' n_{lk}} = a_1$$

where a_1 may be spoken of as the average genotypic excess of the particular gene chosen. Σ is used for summation over allelomorphs of the same factor. If p_1 is the proportion of this kind of gene among all homologous kinds which might occupy the same locus, it is evident that

$$\sum_{k=1}^{s} (p_k a_k) = 0.$$

I would surmise that the above passage is comprehensible to not more than one reader in a hundred and one biologist in ten. Yet it is nothing compared to what Sewall Wright can do:[42]

There may be much more random drift from fluctuations in the selection coefficients. The case of fluctuations in the case of equilibrium maintained by overdominance is of especial interest. Letting $\bar{s}$ and $\bar{t}$ be mean selective disadvantages of the two homozygotes relative to the heterozygotes, we have

$$\overline{\Delta q} = -q(1 - q)[\bar{s}q - \bar{t}(1 - q)]$$
$$= -(\bar{s} + \bar{t})q(1 - q)(q - \hat{q}), \quad \hat{q} = \frac{\bar{t}}{\bar{s} + \bar{t}},$$
$$\delta q = q(1 - q)[s - \bar{s})q - (t - \bar{t})(1 - q)],$$
$$\sigma^2{}_{\Delta q} = q^2(1 - q)^2[\sigma_s^2 q^2 + \sigma_t^2(1 - q)^2 - 2\sigma_s\sigma_t r_{st}q(1 - q)].$$

Consider, first, the case in which there is fluctuation merely in intensity ($r_{st} = 1$, $\hat{q}$ constant):

$$\sigma^2{}_{\Delta q} = \sigma_{s+t}^2 q^2(1 - q)^2(q - \hat{q})^2.$$

Substitution in the formula for $\phi(q)$ gives an expression in which the ordinate at $q = \hat{q}$ is always infinitely greater than at any other value of q, indicating that the distribution is confined to the equilibrium point, as it obviously must be, since $\sigma_{\Delta q}$ is merely a multiple of Δq in this case.

If, however, s and t vary equally and in perfect negative correlation,

$$(\sigma_s^2 = \sigma_t^2, \qquad r_{st} = -1, \qquad s + t \text{ constant}),$$

$$\sigma^2_{\Delta q} = \sigma_t^2 q^2 (1 - q)^2,$$

$$\phi(q) = \frac{\Gamma(a - 2)}{\Gamma(a\hat{q} - 1)\Gamma[a(1 - \hat{q}) - 1]} \; q^{a q - 2}(1 - q)^{a(1 - \hat{q}) - 2},$$

$$a = \frac{2(\bar{s} + \bar{t})}{\sigma_t^2},$$

$$\bar{q} = \frac{a\hat{q} - 1}{a - 2},$$

$$\sigma_q^2 = \frac{\bar{q}(1 - \bar{q})}{a - 1}.$$

These men make biology into what Hardin calls a "numbers game." [43] They have little of the naturalist's burning interest in plants and animals. They are an alien element, and yet the biologists have paid lip service to them.[44] I would like to see the naturalists stand on their own feet and overcome their servility.

I would even suggest that the American Institute of Biological Sciences appoint a commission of competent biologists to review all the work of Fisher, Ford, Haldane, and Wright and decide whether it has added anything whatever to our understanding of nature. This is not as mad as it sounds, because we have been through all this before. Sir Julian Huxley, doubtless aided by his long family memory, reports that Francis Galton (Darwin's cousin) and his disciple Karl Pearson, in the early days, applied mathematical methods of extreme delicacy and ingenuity to the study of evolutionary problems.[45] In vain. They were working from assumptions that

proved to be erroneous, and their labors were useless. Huxley
thinks it is different now because the present mathematicians
are working on "a firm basis of fact," but the day may come
when their facts are seen to be erroneous assumptions and
their labors go into the wastebasket.

1. Simpson (1933), 138.
2. Fisher (1930), 240; Huxley (1953), 34; Mayr (1963), 183, 199; Simpson (1944), 180–181; (1949), 224, 268.
3. Simpson (1953), 138.
4. Huxley (1957), 35; see also Williams (1966), 6–7.
5. Rensch (1960), 3.
6. Simpson (1944), 40.
7. Stebbins (1950), 152.
8. Huxley (1942), 23.
9. Stebbins (1950), 101; a fine modern example of uncontrolled speculation can be found in Morris (1967), Chapter 2.
10. Simpson (1944), 74; (1953), 132.
11. Stebbins (1950), 119.
12. Mayr (1963), 190.
13. Dobzhansky (1941), quoted in Simpson (1953), 88.
14. Simpson (1953), 278.
15. Simpson (1953), 118; see also Huxley (1942), 463–464.
16. Simpson (1953), 118–119.
17. Hardin (1961), 61; Olson (1960), 534.
18. Simpson (1944), 81.
19. Deevey (1967), 635.
20. Goldschmidt (1940), 6–7; Mayr (1961), 1503–1504; Olson (1960), 540–542.
21. Dobzhansky (1956), 339–340.
22. De Beer (1966), 29.
23. Simpson (1964), 276.
24. Simpson (1969), 123.
25. Ardrey (1966), 175.
26. Ardrey (1966), 280.
27. Simpson (1953), 170.
28. Huxley (1960A), 41–43.
29. Simpson (1949), 160.
30. Ardrey (1961), 49, 58, 264, 320.
31. Darwin (1859), 84.
32. Stebbins (1950), 103, 104, 118, 119, 500, 501.

33. Waddington (1960), 385; Lerner (1958), 10, takes much the same position. See also Simpson (1964), 79.

34. Simpson (1944), 180, 181; (1949), 224, 268.

35. Simpson (1953), 59.

36. Simpson (1953), 138.

37. Simpson (1964), 199, condemns precisely this sort of reasoning among his opponents.

38. Simpson (1953), 146.

39. Barzun (1964), 215–216; Hardin (1961), 61; Huxley (1957), 34. In the sixth edition of *The Origin of Species* (Chapter 4, paragraph 2) Darwin himself referred to natural selection as a "metaphorical expression" and also said: "In the literal sense of the word, natural selection is a false term. . . ."

40. Fischer (1970), 243–259, shows the dangers at great length.

41. Fisher (1930), 34.

42. Wright (1960), 461.

43. Hardin (1961), 133.

44. Deevey (1967), 640.

45. Huxley (1942), 151–152; see also Goldschmidt (1940), 137; Hofstadter (1955), 161–167, and Lerner (1958), vii, 2, 15 n. I must also cite Fischer (1970), 287–289, even though it can obviously be used against me.

Darwin did not invent the struggle for existence. As Eiseley points out, it is an "obvious and self-evident fact," [1] and it had been mentioned by naturalists several times before Darwin was born. What Darwin did was to make the phrase a familiar shibboleth, assign a creative role to the process, and praise it as virtuous. In a way that none of my paperback authors would dare to imitate nowadays, he asserted that it favored the welfare of the right sorts:

> All that we can do, is to keep steadily in mind that each organic being is striving to increase at a geometrical ratio; that each at some period of its life, during some season of the year, during each generation or at intervals, has to struggle for life, and to suffer great destruction. When we reflect on this struggle, we may console ourselves with the full belief, that the war of nature is not incessant, that no fear is felt, that death is generally prompt, and that the vigorous, the healthy, and the happy survive and multiply. [2]

Darwin's followers, in their enthusiasm for the principle, carried it to extraordinary lengths. [3] T. H. Huxley said that all the molecules within each organism were competing with each other. August Weismann suggested that the particles of germ plasm were in conflict with each other, so that the ancestors who had contributed them could be seen as struggling with each other as to which should be re-created. Wilhelm Roux developed the theory that the organs were struggling with each other for nourishment, kidneys against lungs, heart against brain. Neither Darwin nor his immediate followers had much feeling for the internal stability and harmony of the organism.[4]

Darwin was not working in a vacuum, but in nineteenth-century England. His ideas, or rather his slogans, were caught up at once and applied in the social sphere. As Simpson says, with much restraint: "These concepts had ethical, ideological, and political repercussions which were, and continue to be, in some cases, unfortunate."[5] G. B. Shaw, using no restraint, gives a more colorful description:

> Never in history, as far as we know, had there been such a determined, richly subsidized, politically organized attempt to persuade the human race that all progress, all prosperity, all salvation, individual and social, depend on an unrestrained conflict for food and money, on the suppression and elimination of the weak by the strong, on Free Trade, Free Contract, Free Competition, Natural Liberty, Laisser-faire: in short, on "doing the other fellow down" with impunity.[6]

When the first enthusiasm wore off and the bill for the damages came in, the biologists realized that things had gone too far. There had been bad science as well as bad sociology, and they had to put their house in order. This was accomplished in two ways.

First, the emphasis on struggle was played down. Instead of being obvious and self-evident, it became almost invisible. Simpson, for example, allows it practically no role in the modern view of evolution:

> Struggle is sometimes involved, but it usually is not, and when it is, it may even work against rather than toward natural selection. Advantage in differential reproduction is usually a peaceful process in which the concept of struggle is really irrele-

vant. It more often involves such things as better integration
into the ecological situation, maintenance of a balance of nature, more efficient utilization of available food, better care of the
young, elimination of intra-group discords (struggles) that
might hamper reproduction, exploitation of environmental possibilities that are not the objects of competition or are less effec-
tively exploited by others.[7]

Second, the influence of cooperation in nature was emphasized. This was not difficult, since cooperation is as obvious
and self-evident in nature as struggle had ever been. In Rus-
sia, even before the Bolshevik Revolution, the scientists always laid more stress on mutual aid than on competition.
Nowadays this is also fashionable in the West.[8] Symbiosis
and ecology are popular. Biologists recoil in horror from
Tennyson's famous line about "Nature red in tooth and
claw." Professor W. C. Allee expresses the modern attitude
when he says: "The . . . life of animals shows two major
tendencies: one towards aggressiveness, which is best developed in man and his fellow vertebrates; the other towards
. . . cooperation. . . . I have long experimented upon both
tendencies. Of these, the drive toward cooperation . . . is the
more elusive and the more important." [9]

It is my belief that Allee represents the general opinion of
the biologists and would be indorsed by most reasonable men.
Darwin himself might well go along. But it is only fair to say
that there are eminent men who deviate from Allee in both
directions.

Sir Julian Huxley, for example, goes even further than
Simpson in toning down the struggle. He makes the following remarkable statement: "The struggle for existence merely
signifies that a portion of each generation is bound to die
before it can reproduce itself." [10] Here Darwin's original concept is utterly denatured. There is no struggle at all. Some die
before maturity, but that sad fact was known to Solomon. It
is a truism. Darwin would not regard it as a discovery.

Hardin, on the other hand, refuses to depart from the early position. He has strong political views and despises the utopians who try to get away from competition.[11] In his eyes, no activity of man—not even painting, sculpture, music, or writing—is without its competitive aspect. Nor does he regard this as a personal idiosyncrasy; he asserts that it is biology: "It is a basic axiom of biology that *the struggle for existence cannot be suppressed;* it can only be altered in the form it takes." [12]

Perhaps the most surprising dissenter is Deevey. As an ecologist, he must be especially aware of the cooperative aspects of nature. He confesses that "the shibboleths of Darwinism—struggle for existence, survival of the fittest—had a shockingly Teutonic tone," and that there must be something wrong with a doctrine that encouraged such vicious distortions. Nevertheless, he declares that these shibboleths "survive today essentially as Darwin propounded them." [13] I am reluctant to assume from this that Deevey has not read Simpson, Allee, and Huxley, or that he really disagrees with them. When he wrote these words he was taking a polemical position, which he might not want to defend if soberly pressed.

With such diverse opinions emanating from Harvard, Yale, and the University of California, it is impossible to say that the biological fraternity takes this or that position. It is, however, obvious and self-evident that the fraternity is no longer solidly in favor of the doctrine of the struggle for existence as propounded by Darwin.

The discerning reader may be distressed to see that the struggle for existence is seldom discussed as a *biological* problem. This has been its history from the beginning. Darwin took it over from Malthus, who was a sociologist (and a grim one) rather than a biologist. It was not derived from a loving contemplation of plants and animals. Such a contemplation would show that there were always more seeds than were needed for the replacement of the parents, but it would not show that "each organic being was *striving* to increase at a geometrical ratio" or that there was continual struggle. Striving and struggle are largely human traits and cannot be imputed to pollen grains or fish eggs. Nature, in her usual ambiguous way, offers examples of strife and other examples of cooperation, and she is not consistent enough to yield a firm basis for a theory.

In order to conclude this chapter on a biological note, let me mention a curious phenomenon that is reported several times by Simpson; is known to, but seldom mentioned by, his colleagues; and is utterly unknown to the general public. This is deferred replacement.[14] Porpoises and dolphins, for example, have replaced ichthyosaurs in their "adaptive zone," but it must not be assumed that they accomplished this by struggle or competition. The ichthyosaurs became extinct long before the porpoises and dolphins appeared, and during the interval the adaptive zone was simply empty. There was never any confrontation or battle. Such cases, according to Simpson, are numerous.

This is a truly biological fact, derived from a perceptive study of the fossil record. It is entirely foreign to the common belief, disseminated by many popularizers and taken as a matter of course by most laymen, that the history of the earth has been one long bloody fight. A passage such as the following (by two outsiders) is bad biology: "At night and in winter, as the great reptiles lay torpid, the mammals took over and the reptiles were driven into oblivion." [15] The professionals seldom say anything as crass as this, though Olson reports a "popular theory" that the mammals became egg-feeders and ate many reptilian eggs.[16] Inevitably I sometimes doubt that the implications of deferred replacement, especially in its bearing on the struggle for existence, have been fully digested.

1. Eiseley (1961), 52.
2. Darwin (1859), 78–79.
3. Eiseley (1961), 334–336.
4. Eiseley (1961), 336.
5. Simpson (1949), 221; see Hofstadter (1955) for a fuller treatment.
6. Shaw, Preface to *Back to Methuselah*.
7. Simpson (1949), 222.
8. Huxley (1942), 479–480.
9. Allee (1955), 243.
10. Huxley (1957), 34.
11. If the reader wonders what Hardin means by competition, he will have good company. Mayr (1963), 86, remarks that "our ignorance of the nature, the mechanisms, and the amount of competition is vast."
12. Hardin (1961), 218–220.
13. Deevey (1967), 635, 640.
14. Simpson (1949), 118–119, 248.
15. Gamow and Ycas (1968), 149.
16. Olson (1965), 130.

The phrase "survival of the fittest" was not coined by Darwin. He took it over from Herbert Spencer, apparently considering it an improvement on his own natural selection.[1] It immediately became an integral part of classical Darwinism, much to the embarrassment of modern adherents.

Survival of the fittest has suffered the same blight as its companion shibboleth, struggle-for-existence. It is politically unacceptable. It smells of Hitler, of the laissez-faire economists, of savage competition and devil take the hindmost. The biologists, sorry that it was ever mentioned, do their best to forget it. Smith, Huxley, and Eiseley say not a word about it. Hardin, whose political passions sometimes warp his scientific judgment, mentions it briefly, but even he puts quotation marks around the word *fittest*.[2] I had to look far beyond my paperbacks to find out what had happened.

I discovered that the phrase had been discredited long before the political blight descended upon it.[3] Very early, so early that I cannot ascertain the date, someone asked how we determine who are the fittest. The answer came back that we determine this by the test of survival; there is no other criterion. But this means that a species survives because it is the fittest and is the fittest because it survives, which is circular reasoning and equivalent to saying that whatever is, is fit. The gist is that some survive and some die, but we knew this at the outset. Nothing has been explained.

The late J. B. S. Haldane, despite his Marxist leanings, was fully accepted as a member of the prevailing synthetic school of evolutionists.[4] For this reason, and especially because he was a mathematician, I was surprised that he should recognize survival of the fittest as a tautology, but still not object to it as such. He said: ". . . the phrase, 'survival of the fittest,' is something of a tautology. So are most mathematical theorems. There is no harm in stating the same truth in two different ways."[5] This is extremely misleading. There is indeed no harm in stating the same truth in two different ways, if one shows what one is doing by connecting the two statements with a phrase such as *in other words*. But if one connects them with *because,* which is the earmark of the tautology, one deceives either the reader or oneself or both; and there is ample harm in this. The simplest case, where one is informed that a cat is black because it is black, may be harmless, though irritating and useless; but the actual cases are always harder to detect than this, and may darken counsel for a long time. I cannot believe that Haldane is doing justice to most mathematical theorems, since the connector there is not so much *because* as *behold.**

It has never been possible to break out of the circle by finding a better word than *fittest*. But, since something had to be done to restore logical respectability, a new meaning was

* There are surprisingly few systematic discussions of fallacies. As nearly as I can make out, a tautology is bad because it employs the same term on both sides of the equation; e.g., your deafness is caused by the impairment of your hearing. The element of causation seems to be necessary and there is, of course, always a masking of the identity on one or both sides. Circular reasoning is bad because it asserts that A is caused by B and B is caused by A. The error is in the reasoning about causation rather than in the confusion of identities. The two fallacies are not carefully distinguished in common practice, and Haldane probably regarded them as one and the same.

foisted on the old word. *Fitness* was redefined to mean "having the most offspring." Mayr says: ". . . those individuals that have the most offspring are by definition . . . the fittest ones." [6] Deevey echoes him: "Let us also rephrase *survival of the fittest* as *differential survival,* clearing out a word that begs a huge question." [7]

Simpson, the dean of the evolutionists, nails the point down even more firmly, stating that among geneticists fitness has nothing to do with the common understanding of the term: "If genetically red-haired parents have, on an average, a larger proportion of children than blondes or brunettes, then evolution will be in the direction of red hair. If genetically left-handed parents have more children, evolution will be toward left-handedness. The characteristics themselves do not directly matter at all. All that matters is who leaves more descendants over the generations. Natural Selection favors fitness only if you define fitness as leaving more descendants. In fact geneticists do define it that way, which may be confusing to others. To a geneticist fitness has nothing to do with health, strength, good looks, or anything but effectiveness in breeding." [8]

Thus robbed of its normal and customary meaning, I did not expect the word *fit* to appear any more in the professional literature. Professor Mayr apparently agreed, since he condemned "such trivial and meaningless circular statements as, for instance, *the fitter individuals will on the average leave more offspring.*" [9] But old habits die hard. Even with his own words and Mayr's before his mind, Simpson lapsed into saying: "On an average, more offspring will survive from those parents whose heritable variations make them more fit." [10] If we adopt Simpson's own definition of the word *fit,* and if we remove all surplus language, this sentence can be rewritten thus: "On an average, more offspring will survive from those parents who leave more offspring." Simpson is too intelligent to say anything like this purposely, but he is struggling with a treacherous set of words.

The reader will recall that natural selection, like survival of the fittest, has been redefined to mean differential reproduction or leaving the most offspring.* He will also recall that differential reproduction, as an explanation of evolution, boils down to a meaningless tautology. Thus the effort to salvage some meaning for survival of the fittest has produced nothing. In trying to break out of the circle, the scholars only fell into a tautology.

So much for the concept of survival of the fittest. We must now add that, although the word *fit* is generally shunned, something is needed to replace it as an adjective. The choice has fallen on the rather colorless word *adaptive,* which is now widely used to signify that something is useful or advantageous. Thus if the horse acquires teeth of a new style, and if these chew grass better than the old ones did, the change is adaptive. Conversely, if the Irish elk acquires antlers so enormous that it can barely carry them, the change may be suspected to be inadaptive, i.e., injurious.[11]

The sympathetic observer will be disappointed to learn that this does not cure the troubles. The word *adaptive* sounds harmless and has not led to political entanglements, but it is unmanageable in practice. No one can decide, with the naked eye or with instruments or with mathematics, whether a given trait is adaptive, inadaptive, or neutral. Bateson recognized this as early as 1894.[12] Simpson asserts it as indisputable: "In the nature of things it is quite impossible to establish that every

---

* Darwin apparently regarded natural selection and survival of the fittest as different ideas in 1859; at least, he speaks of them that way in Chapter 4 of the first edition of *The Origin of Species.* But when he prepared the sixth and last edition in 1872, he must have come to see them as the same thing. Chapter 4 is now entitled "Natural Selection, or the Survival of the Fittest," and a new second paragraph is inserted in which Darwin says that survival of the fittest is a "more accurate" expression of what he had previously called natural selection.

single genetic difference between two populations has selective value, and probably some distinctions differ in this respect; but neither is it possible to prove that they are really indifferent." [13] Since having selective value is the same thing as being adaptive, this comes down to saying that we do not know what is and what is not adaptive. Can a solid theory be built on such a base?

There have been students who denied that any traits could be neutral, but Mayr has no doubts about this. He announces firmly that the presence of spots means only that spots are present: "If a given subspecies of ladybird beetles has more spots on the elytra than another subspecies, it does not necessarily mean that the extra spots are essential for survival in the range of that subspecies. It merely means that the genotype that has evolved in this area as the result of selection develops additional spots on the elytra." [14]

Another strange aspect of the fitness problem is the vast gulf it reveals between the biologists and the ordinary laymen. The biologists have discarded survival of the fittest, together with the words *fit* and *fitness* in their normal usage, and they have been perfectly open about this.[15] Yet if my acquaintances are typical, the laymen have noticed nothing. They continue to talk of fitness in the old way and to regard survival of the fittest as sound doctrine.

1. Barzun (1958), 76.
2. Hardin (1961), 57.
3. Bateson (1894), 79–80; Mayr (1963), 183; Nicholson (1960), 479.
4. Simpson (1949), 278.
5. Haldane (1935), 24.
6. Mayr (1963), 183.
7. Deevey (1967), 635.
8. Simpson (1964), 273.
9. Mayr (1961), 1504.
10. Simpson (1964), 51.
11. Simpson would not agree, being convinced that he can explain away all apparent inadaptivity: (1953), 282–291. He admits that Haldane, Waddington, and Huxley disagree with him.
12. Bateson (1894), 10–13.
13. Simpson (1944), 78.
14. Mayr (1963), 311.
15. McAtee (1937).

Biologists for the most part look and talk like prosaic men, but many of them became biologists because they were fascinated by the wonders of nature, especially the extraordinary complexities, adjustments, and inventions that are commonly spoken of as "adaptations." Who can fail to be impressed by these things and to admire the patience and diligence of the field workers who discovered and described them? They are the glory of the profession.

Darwin himself was keenly aware of these wonders and departed from his habitual sobriety to call them "exquisite" and "beautiful."[1] Sir Julian Huxley, also a sober man, is equally charmed: ". . . every plant and every animal is . . . an organized bundle of adaptations—of structure, physiology, and behavior; and the organization of the whole bundle is itself an adaptation."[2] My other paperback authors are not behindhand, nor are the critics and skeptics. All unite in wonder at the works of nature.

This happy unanimity dissolves as soon as interpretation and explanation begin. Bitter controversies rage over what the demonstrated facts signify, how they have come about, and why they are as they are.

Let me start by disposing of an unfortunate semantic problem. For one reason or another, there has been a tendency to equate adaptation with fitness and survival (which have already been equated with each other). It is obvious, of course, that there must be a certain amount of harmony between organism and environment; a fish will die without water and a bird will die without air. But this does not mean that every living species is well adapted or that every extinct species was ill adapted. Simpson warns us of the danger of using the word in this sense;[3] we would be arguing that species die out because they lose adaptation, while at the same time concluding that they have lost adaptation because they have died out. This is the same problem that we encounter in survival-of-the-fittest, where we fall into circular reasoning by saying that we survive because we are fit and are fit because we survive. That fallacy crops up repeatedly, and so does this one. Simpson himself frequently succumbs to it despite his own warning. Thus he remarks: "The primitive ameba has remained adapted, hence has survived, while the lordly dinosaur lost adaptation and therefore life."[4] By Simpson's own showing, this means only that the ameba has survived while the dinosaur has died out, which is correct but not newsworthy. The air will be clearer if we look at adaptation without a side-glance at survival.

Even if all individuals and species are to some extent adapted, there are different degrees and sorts of adaptation.[5] I have gradually come to think that there are three general types, with all kinds of intergrading between them. My examples will be taken from the animal realm, since most of the discussion is by zoologists.

First, many animals seem to be poorly adapted. Thus the gorilla, though supposedly designed for swinging from bough to bough, scrounges for a living on the earth and rarely climbs a tree.[6] Several features of the human body do not seem to be well designed for our way of life; e.g., the vermiform appendix, the ear-wiggling muscles, and the valves in the horizontal blood vessels that run between the ribs (where they are not needed).[7]

The most famous examples are the enormous antlers of the Irish elk and the ponderous tusks of the mammoth, organs that had many disadvantages and no visible utility. These animals are now extinct, which gives rise to a terrible temptation to ascribe their extinction to bad adaptation, thereby entering the circle mentioned above. This temptation must be resisted because there is no evidence as to the exact cause of extinction (they may all have been drowned in a flood so far as we know) and also because, as Professor Simpson shrewdly remarks, when the animals were abundant for tens of thousands of years despite their burdens, it is hard to be sure that the structures were disadvantageous.[8] We must simply recognize that animals can exist for ages even if, to our eyes, they are very poorly adapted.

Second, many animals have a talent for camouflage and mimicry. Sometimes their achievements are trivial, as in the oft-quoted phenomenon of industrial melanism, where certain moths take on a dark coloring where industry has sprinkled the trees with soot. But sometimes these are amazingly complex, as in the insects that mimic dead leaves. The illusion can be heightened by the appearance of holes, and accordingly we find that in some cases there are genuine holes in the wings, but in others a hole is suggested by the absence of scales over a part of the wing, while in still others the shading and coloration are such as to simulate a hole.[9] The subject-object relationship seems clear, and there is obvious utility in the performance, hence these cases are, to my mind, adaptations par excellence.

Third, many animals behave in incredibly complicated and mysterious ways. The life course of the swallowtail butterfly is a familiar example (though few laymen know the remarkable details, such as the total dissolution and reconstitution of the organs and even of the cells), but other insects are equally striking. These are marvels, beyond any doubt; but there is no compelling reason to regard them as adaptations. Each is a tour de force by a virtuoso, but the virtuoso seems to be exercising his own fantasy rather than adapting himself to mundane conditions in a utilitarian way.

The books are full of examples of this virtuoso work, which is especially common among insects. I will set out one case at length to show how many refinements there can be and how the whole performance shows a master hand.

In early summer the small wasplike *Eumenes amedei* of northern Africa and southern Europe emerges from the pupal state as an elegant insect with yellow and black bands. Soon after mating, the female prepares a house in which her young can develop and sufficient food can be stored. She chooses an exposed and sunny situation on a rock or wall, and builds a circular fence of small stones and mortar, the mortar being made from dry flinty dust mixed with her own saliva. The stones are chosen with care, flint being preferred to limestone, and the fragments selected are all much the same size. Her choice of the most polished quartz fragments suggests (if we are anthropomorphic) that she is not indifferent to the esthetic effect of her handiwork. As the wall grows higher, the builder slopes it toward the center and so makes a dome which, when finished, is about the size of a small cherry. A hole is left at the top, and on this is built a funneled mouthpiece of cement.

The next task is to collect the food supply for the future grub. This consists of small caterpillars about half an inch long, palish green, and covered with white hairs. These caterpillars are partially paralyzed by the sting of the *Eumenes* and are unable to make any violent effort to escape. They are stored on the floor of the cell. Since they remain alive, they keep fresh until the grub is ready to eat them; if they were killed outright, their flesh would soon dry up or rot. When the cell is stocked, a single egg is laid in each house, and the mouthpiece at the top of the cell is closed with a cement plug, into which a pebble is set.

The egg is not laid upon or among the caterpillars, as in many allied species. These caterpillars are only partially paralyzed, and can still move their claws and champ their jaws. Should one of them feel the nibblings of the tiny grub, it might writhe about and injure the grub. Both the egg and the grub must be protected, and to this tend the egg is suspended by a tiny thread of silk fastened to the roof. The caterpillars may wriggle and writhe, but they cannot come near it.

When the grub emerges from the egg, it devours its eggshell, then spins for itself a tiny silken ribbon-sheath in which it is enfolded tail-uppermost and with head hanging down. In this retreat it is suspended above the pile of living food. It can lower itself far enough to nibble at the caterpillars. If they stir too violently, it can withdraw into its silken sheath, wait until the commotion has subsided, then descend again to its meal. As the grub grows in size and strength, it becomes bolder; the silken retreat is no longer required; it can venture down and live at its ease among the remains of its food.

The stone cells are not all stored with the same wealth of caterpillars. Some contain five and some ten. The young females, larger than the males, need twice as much food. But note that the cells are stocked before the eggs are laid, and that biologists generally believe that the sex is already determined when an egg is laid. How does the *Eumenes* know the future sex of her eggs? How is it that she never makes a mistake?

To characterize the three types a little further, let us refer to them as clowns, craftsmen, and wizards.[10] The clowns do a poor job of fitting into nature. The craftsmen, with their utilitarian ingenuity, fit themselves in with consummate art. The wizards use nature as their clay, but are not so much fitting into it as rising above it. The three styles are so different that the word "adaptation" (or probably any other word) does not fit them all happily; they are unequal yoke-fellows, as the old grammarians used to say. Nevertheless, the usage is now so well established that the peculiar manner of life of each group must be described as its adaptation.*

* The term is also unfortunate in tending to assume one of the main points in question, viz. alteration under the influence of natural selection. Therefore the reader must be warned that most of these marvels have no pedigree,

How do these different groups fare in nature? Strange to say, they all seem to get along in much the same way. The clowns do not die out, even if the mammoth and the Irish Elk are gone.[11] The craftsmen do not take over the earth. The wizards maintain their places with no apparent gain or loss. Darwin observed that the numbers of a given species actually remain more or less constant, and this is confirmed by later students.[12] It is possible to infer from this that crafty adaptation is not really a matter of life or death, but little attention is paid to this line of thought.† La Fontaine's fable of the ant and the grasshopper is very impressive for children, but the grasshoppers do not die out despite their improvidence.

We must soon approach the battlefield, but one further consideration will enable us to see the controversies in better perspective and to realize the value of our threefold classification. The attentive student soon notices that no one pays much attention to the clowns; indeed, they are hardly ever mentioned. The Darwinians, who have always had a strong utilitarian bias,‡ stress the work of the craftsmen as much as possible, harping endlessly on industrial melanism. The anti-Darwinians (of practically all stripes and colors) stress the work of the wizards, rejoicing in the lack of utilitarian value and in the difficulty of explaining the magic.

———

meaning that there is no record of gradual development from earlier and less marvelous arrangements: Gray (1876), 214, 319; Darwin (1871), 761.

† McAtee (1932) is the chief article I have found on this point. Analyzing reports on eighty thousand bird stomachs, he concluded that animals were eaten by birds pretty much in proportion to their availability, and that "protective adaptations" were of no advantage. He was rather harsh (2–3) about his colleagues' experimental methods and attempted explanations, saying: "Undeniably selectionists have been absurd in their disquisitions on adaptations."

‡ Darwin himself, in his later years, confessed that this was a flaw in his work: "I did not formerly consider sufficiently the existence of structures which, as far as we can . . . judge, are neither beneficial nor injurious, and this I believe to be one of the greatest oversights as yet detected in my work. This led to my tacit assumption that every detail of structure was of some special though unrecognized service." (Eiseley, 1969, 142.)

Now let us look at the ancient contention as to the significance of adaptations. The achievements of the craftsmen and
the wizards were well known by 1850, and were commonly
cited to the public as showing the wonderful handiwork of the
Creator. When Darwin came forward with a theory that excluded the Creator, he was immediately challenged to explain
these achievements by his method of slow step-by-step changes.
He wrestled manfully with this task, especially with the problem of the human eye;* he suggested lines of thought that
might be fruitful for later investigation; and he made some
illuminating remarks about the arguments pro and con.[13] But
not even his warmest admirers would say that he had met the
challenge. Nor have his followers. Goldschmidt (a purely scientific critic with no religious motives) was able to say in 1940
that the eye and sixteen other important features remained un-
explained on the strict Darwinian view of accumulation and
selection of small mutations.[14]

The literature on this subject is meager, despite its importance to evolutionary theory.[15] There are many general decla-
rations of what evolution must have been like, but very few
studies of specific cases. This is no accident; the biologists have
become wary from bitter experience. They remember that they
were carried away by their so-called explanations in the early
years and they know that the dangers are still present.[16] I will
illustrate this as to both how and why.

The efforts to explain why led to absurdities. It is easy to say
that legs are for running and ears are for hearing, but trouble
begins when one ventures a little further. For instance, Sir
Julian Huxley says: "Flowers develop distinctive colors to

---

* This problem appears frequently in the literature, although the placenta
and various other organs might do as well. We will recur to it often. The
reader should be informed, however, that the evolution of the eye in man
(and in all the vertebrates) is a major mystery; and that, small as it is,
the eye is an enormously complex structure of retina, cornea, rods and
cones, visual purple, muscles, nerves, and fluids. Supporters of natural
selection tend to play down this complexity, while opponents emphasize it.

attract bees; wasps develop their black and yellow stripes to
warn enemies of their stings; the partridge develops camouflage to escape detection by the hawk; the peacock develops
brilliant plumage to stimulate his mate." [17] This sort of interpretation is simplistic and open to considerable doubt as to its
correctness. It is also open to embarrassing questions, such as
why the well-camouflaged grasshopper betrays his location by
chirping. But at least it is not arrant nonsense such as we find
in Professor Tinbergen of Oxford (second only to Konrad
Lorenz in the new science of ethology or animal behavior)
when he says: ". . . the brightly colored patches of skin
seen round the genital aperture of female Baboons and
Chimpanzees probably guide the male to the female's copu-
latory organs." [18] This kind of childish "explanation," which
offhandedly assumes that baboons and chimpanzees need
more guidance than other primates, is precisely what brought
Darwinism into contempt around the turn of the century.[19]
Obviously it is still with us today, although most modern biologists are too sophisticated or too wary to fall into such errors.
They have learned that it is not wise to try to explain why.*

The answer to how is equally difficult, and one seldom finds
a scientist rash enough to make an attempt. We are told that
the miracles were accomplished by natural selection working
in its usual step-by-step manner, but the steps are not shown.
Here is an example of what I call the wave-the-wand method
of explanation, although admittedly it is culled from a popular book by two professors who are not among the elite of the
evolutionists:

> The animals changed too. Some of the reptiles in the colder
> regions began to develop a method of keeping their bodies
> warm. Their heat output increased when it was cold and their
> heat loss was cut down when scales became smaller and more

* A glaring exception to this cautious attitude may be found in Chapter 2
of *The Naked Ape* (Dell, 1969), where Desmond Morris explains various
aspects of human sexual organs and behavior by pure imagination.

pointed, and evolved into fur. Sweating was also an adaptation to regulate the body temperature, a device to cool the body when necessary by evaporation of water. But incidentally the young of these reptiles began to lick the sweat of the mother for nourishment. Certain sweat glands began to secrete a richer and richer secretion, which eventually became milk. Thus the young of these early mammals had a better start in life.[20]

I have never found any responsible evolutionist publicly creating out of whole cloth in this way. The professionals, being more prudent, do not try to explain. Thus Simpson, in one of his latest books, declares that the days of confusion are over and that the synthetic theory has found the answers, but never does this sagacious man venture to discuss a specific case or to spell out the infinite number of small steps needed.[21] The situation is made even clearer in an article written by Mayr for the Darwin Centennial Celebration in Chicago in 1959, an occasion when great efforts were made to praise and justify the Founder.[22] This article, entitled "The Emergence of Evolutionary Novelties," was read in draft form by W. Bock, Julian Huxley, B. Patterson, G. G. Simpson, and C. H. Waddington, all of whom made valuable suggestions; hence it should represent the best thinking of the profession. Mayr is a learned man and has perused an immense amount of literature, but he confines himself to the usual general remarks and does not pretend to explain any specific problem such as the intractable human eye. He too is sagacious.

The reader will be interested to know that Darwin himself was not so prudent. He was willing to stake everything on meeting the challenge of the marvels. He actually said: "If it could be demonstrated that any complex organ existed, which could not possibly have been formed by numerous, successive, slight modifications, my theory would absolutely break down." [23] Since this fact seems to have been demonstrated, if only by default, the reader will ask whether the modern Darwinians concede that the theory has broken down. The answer is a strange one—they are not greatly troubled by their failure to explain the adaptations because they are sustained and soothed by the best-in-field fallacy.

Darwinism has had to compete with various rival theories, each of which aimed to be a more or less complete explanation. The most famous rivals were vitalism, fundamentalism, Lamarckism, and the hopeful-monster suggestion of Goldschmidt. The Darwinians have shown that none of these theories are any good. Simpson can shoot down each and every one of them with ease. Thus the Darwinians are able to say that Darwin made a better try than anyone else, and they find real comfort in this.[24]

Does this mean that Darwinism is correct? No. Sir Julian Huxley says that, once the hypothesis of special creation is ruled out, adaptation can only be ascribed to natural selection, but this is utterly unjustified.[25] He should say only that Darwinism is better than the others. But when the others are no good, this is faint praise. Is there any glory in outrunning a cripple in a foot race? Being best-in-field means nothing if the field is made up of fumblers.[26]

When the most learned evolutionists can give neither the how nor the why, the marvels seem to show that adaptation is inexplicable. Yet those who cannot explain it will not admit that it is inexplicable. This is a strange situation, only partly ascribable to the rather unscientific conviction that evidence will be found in the future.[27] It is due to a psychological quirk that Simpson describes with admirable self-knowledge: "For some, adaptation was merely an inexplicable fact; these students were few, because scientists rarely are psychologically capable of accepting a phenomenon as a fact and also accepting it as inexplicable." [28] This observation may be correct, but it is not to the credit of the profession. Scientists are expected to rise above such frailties.

The best-in-field fallacy seems to be my own discovery. It does not appear in books on fallacies and I have not seen it clearly expressed anywhere else. Perhaps it appears with unusual frequency among the evolutionary theorists, who seem to have a special weakness for it.

My best example comes from Mayr, although he is normally a highly intelligent man. In the passage concerned he concedes that there are valid objections to his theory, but he rules out these objections on the ground that their proponents have not advanced a better suggestion: ". . . it is a considerable strain on one's credulity to assume that finely balanced systems such as certain sense organs (the eye of vertebrates, or the bird's feather) could be improved by random mutations. This is even more true for some of the ecological chain relationships (the famous yucca moth case, and so forth). However, the objectors to random mutations have so far been unable to advance any alternative explanation that was supported by substantial evidence." [29]

It seems that the standards of the evolutionary theorists are relative or comparative rather than absolute. If such a theorist makes a suggestion that is better than other suggestions, or better than nothing, he feels that he has accomplished something even if his suggestion will obviously not hold water. He does not believe that he must meet any objective standards of logic, reason, or probability. This is a curious state of affairs, but if the reader (allowing for my pride of authorship) can view it as a possibility he will feel less surprise in the frequent cases where he finds the theorists propounding ideas of striking frailty.

This will also help the reader to understand why I do not use what is sometimes called the comparative method, whereby each problem is discussed in the light of Darwinism, Lamarckism, vitalism, and other possible theories. These other theories having long since been rejected, there is no point in discussing them further. Such a discussion would only give a specious satisfaction to the supporters of whichever view was adjudged preeminent among the cripples.

1. Darwin (1859), 60–61.
2. Huxley (1957), 15.
3. Simpson (1953), 294–295; see also Gray (1876), 214.
4. Simpson (1964), 21; see also (1953), 159, 302, and Huxley (1942), 443.
5. Simpson (1953), 161.
6. Ardrey (1961), 112–113.
7. Hardin (1961), 59.
8. Simpson (1953), 286; see also (1949), 202.
9. Smith (1958), 17.
10. Huxley (1942), 443–449, hints at such a grouping.
11. Simpson (1949), 202, shows how clowns persist for ages despite being bizarre, queer, and fantastic.
12. Huxley (1942), 14–15; Bateson (1894), 10–13.
13. Darwin (1859), 186–194.
14. Goldschmidt (1940), 6–7, 215–216, 240, 288–291, 386.
15. Mayr (1960), 348–380. Grant (1963) offers 570 pages on "The Origin of Adaptations," but with remarkable blindness or blandness he steers away from controversy on fundamentals. Williams (1966), though critical of many orthodox positions, also leaves the marvels alone.
16. Huxley (1942), 23; McAtee (1932), 1–3; Stebbins (1950), 101.
17. Huxley (1957), 79.
18. Tinbergen (1954), 234.
19. Stebbins (1950), 101–104; Simpson (1964), 202–205.
20. Gamow and Ycas (1968), 149.
21. Simpson (1964), 202–212. Olson (1960), 530, bewails the lack of specific explanations.
22. Mayr (1960), 349–380.
23. Darwin (1859), 189.
24. Fisher (1930), 21; Hardin (1961), 59; Huxley (1957), 40–41; Simpson (1949), 132, 145, 159, 174–175, 200–202, 247; (1953), 134–135; (1964), 199, 205.
25. Huxley (1942), 430, 478.
26. Gray (1876), 105; Haldane (1935), 29; and Kuhn (1962),

23, all seem to have perceived the best-in-field fallacy, though they have not formulated it clearly.

27. Eiseley (1961), 320; (1958), 119; Fisher (1954), 97.
28. Simpson (1964), 204; see also Kuhn (1962), 77.
29. Mayr (1942), 296. Ghiselin (1969) is more prolific; see 215–216, 218, 221, 227–229.

Darwin was keenly interested in sexual selection. One of his major works, *The Descent of Man and Selection in Relation to Sex*, was largely devoted to it. The term seems to have included, in his mind, cases where females exercised some sort of choice after a display or contest among the males, as well as cases where the males fought among themselves and the female was absent or passive.[1] Since this sort of activity is impossible in the plant world and unknown in large parts of the animal world, Darwin could not have regarded it as a general explanation or mechanism of evolution. He only hoped, with some justification, that it would explain specific cases, such as complicated displays and brilliant plumage among the birds. But even in this modest role it has been a disappointment.

There are cases where selection seems so clear that anyone would share Darwin's optimism. Here is an example, taken from a study of the breeding habits of about eight hundred sage grouse in Wyoming: "After the males had sorted themselves out on the strutting-ground, the hens gathered at five mating spots each the size of a room. Dominance established 1% of the males as what Allee terms master-cocks, 2% as sub-cocks. Copulation occurred *only* at the invitation of the hen; in other words, female prerogative of choice was the next step in natural selection. And the result of that selection was that 74% of all matings were with master-cocks, 1% of the total male population; and 13% with sub-cocks, representing 2% of the males. Rank order of dominance had insured that 87% of that season's crop of young sage grouse be fathered by only 3% of the male population." [2]

It is possible to ask some embarrassing questions even about such a case as this; e.g., how can we tell what qualities the hens were looking for? and how effective is this in altering the form of the species? But it is not necessary to press these points, because the whole line of inquiry bogs down when we examine a number of other cases. In fact, the whims and caprices of nature frustrate all efforts to generalize.

In many species the males go through elaborate dances and display which, to our anthropomorphic minds, can only be competitions for the favor of the hens; but when we observe carefully, we find that the hens are absent, not watching, or busy pecking at food. In other cases gorgeous feathers are displayed to hens who seem to be color-blind. There are even species where the hens mate with the *defeated* cocks as readily as with the victors.[3] The hens do not seem to be anthropomorphic.

Robert Ardrey shows at great length that fighting among animals of the same species is generally concerned with territory rather than with females.[4] One could, of course, still argue that fighting was an important element in natural selection even if Darwin misread the motives of the fighters, but this would be vulnerable to the same questions about just which characteristics are favored by the fighting and just how this favoritism will improve the species. Again, however, the whole inquiry is subjected to a *reductio ad absurdum* by, of all creatures, the cuckoo. The fighting among male cuckoos is not for territory, since they make no nests and have no territory; and is not for the females, since the females are promiscuously shared among the males. The fighting apparently has no meaning whatever. This is how Ardrey, always the dramatist, describes it: "Somewhere above a vast oak tree beside some English field two cuckoos fight for exclusive domain. Neither will homestead his territory, for the cuckoo is parasitical and builds no nest. Neither will use his conquest toward romantic ends, for the cuckoo is polyandrous, and these embattled males, when the fighting is done and the real estate properly apportioned, will amicably share their lovelorn bride. They compete, simply, because they must. They compete for reasons of ancient law, stern and abiding, forgotten by men and cuckoos." [5]

Certain biologists of the older generation, such as Sir Julian Huxley, insist that Darwin had something important in sexual selection.[6] Younger men seem to shun the subject as demanding laborious research with little prospect of valuable results. Thus Smith says: "Darwin's ideas on sexual selection have received little attention from later biologists. In no case has it been demonstrated that such selection occurs in a wild population; this is perhaps not surprising, since it would be necessary to show, not only that the females are selecting as mates some kinds of males in preference to others, but also that, by so choosing, females are increasing the average number of offspring they leave." [7] He seems to be unimpressed even by the case of the sage grouse.

There is another objection which, to my mind, is more fundamental than any yet mentioned. In Chapter 4 we saw that the results of breeding are all hopelessly micro because the small changes do not cumulate indefinitely. When a certain point is reached, the animals thwart the breeders by becoming sterile or reverting to type. I see no reason to believe that the hens, even if they are attentive and discriminating as among the sage grouse, can do any better than the breeders.

But the problem goes still deeper. Even if there is such a process as sexual selection (which is arguable) and even if it produces the structures and behavior in question (which is very doubtful), what it has really brought forth is a monumental challenge to natural selection, the keystone of the whole Darwinian theory. In the peacock and the Argus pheasant (favorite subjects of discussion in this field), we have conspicuous and appetizing animals that cannot run, fly, fight, or hide. As Sir Julian Huxley says: ". . . the display-characters may even be clearly disadvantageous to the individual in all aspects of existence other than the reproductive, as in the train of the peacock, the wings of the argus pheasant, or the plumes of some birds of paradise."[8] By all reasonable standards (and who can really cleave to the doctrine that we should not set up standards or assume to judge fitness?[9]) natural selection should never have allowed such animals to come into existence. But they have not only come into existence, they have stayed there and have not become extinct. Have the birds, through their patterns of sexual choice, established a system in which the race is not to the swift and the battle is not to the strong? If so, they have shaken the whole structure of Darwinism.

The biologists have, of course, perceived this. Fisher and Huxley, for instance, are very much aware of the problem and quite unable to solve it. They can only talk about compromise, equilibrium, and setting up checks on the process.[10] No wonder the younger men leave it alone.

Thus we may have understated the case when we said that sexual selection has been a disappointment. It has not only failed to solve the problems to which Darwin applied it; it has called attention to a glaring weakness in natural selection. It has emphasized the existence of things which, under a reasonable view of that theory, simply cannot be.

1. Darwin (1859), 87–90, 468.
2. Ardrey (1961), 94–95.
3. Matthiessen (1967), 51; see also Darwin (1871), 703–704.
4. Ardrey (1966).
5. Ardrey (1961), 159.
6. Huxley (1938) and (1960B), 13; on the other hand, Kellogg (1925), 134–135, had given up on sexual selection as early as 1925.
7. Smith (1958), 148.
8. Huxley (1942), 427.
9. Dobzhansky (1941), quoted in Simpson (1953), 88; Hardin (1961), 57; Mayr (1963), 190; Simpson (1953), 166, 278, 282.
10. Fisher (1930), 152; Huxley (1942), 427, 484; (1953), 44–46; and (1960B), 13.

We have already discussed adaptation and seen how the marvels in nature defy explanation, but the subject is far from exhausted. We must show how adaptation is tied to probability, how much history there is in this connection, what a part it has played in the debates, and how the evolutionists have attempted to eliminate all difficulties by introducing an unbiological stratagem. The whole argument about Darwinism (as the small and difficult part of evolution) is here epitomized.

Let us begin with an Anglican clergyman, Dr. William Paley (1743–1805), archdeacon of Carlisle. Paley was not a scientist, but he was a good compiler and reasoner. In 1802 he published a book called *Natural Theology, or Evidence of the Existence and Attributes of the Deity Collected from the Appearances of Nature,* as to which Darwin said: "I do not think I hardly ever admired a book more than Paley's *Natural Theology.* I could almost formerly have said it by heart." [1]

Paley saw the hand of the Creator everywhere in nature. The marvels were examples of what the Deity could do. Adaptation (in the sense of marvels) was, he thought, the one unanswerable argument in favor of his view. [2] When Darwin tried to explain the marvels without employing a Creator, he was looking straight at Paley; or rather he was looking at Paley's watch, because the controversy is frequently referred to under this head. It is also referred to as Design *v.* Chance, although the evolutionists dislike the latter term.

In order to present the arguments succinctly, I have composed a dialogue between Paley and Darwin. This is imaginary and anachronistic, since Paley died before Darwin was born; and it is also a composite, since I use the words of modern followers (especially Hardin and Huxley) rather than of the original figures. But, since the thrusts and parries are now familiar and stylized, no violence is done to the views of either man.

Paley: Let us dispense with all preliminary considerations, Mr. Darwin, and begin with a question which, in my opinion, will show us exactly where we stand. If, as seems inevitable to me, you give an affirmative answer, we will be in fundamental Christian harmony. If your answer is negative, there will be much more to discuss.

If you found a watch, full of mechanisms exquisitely adapted to produce a series of operations all leading to the fulfillment of the one central purpose of measuring for mankind the march of the day and night, could you believe that it was not the work of a cunning artificer who had designed and contrived it all to that end? And here is a far more wonderful thing than a watch, a man with all his organs ingeniously contrived, cords and levers, girders and kingposts, circulating systems of pipes and valves, dialysing membranes, chemical retorts, carburetors, ventilators, inlets and outlets, telephone transmitters in his ears, light recorders and lenses in his eyes; is it conceivable that this is the work of chance? that no artificer has wrought here? that there is no purpose in this, no design, no guiding intelligence? [3]

Darwin: Dr. Paley, your watch has haunted me for years, but I have slowly reached the conviction that the answer must be negative. I do not believe that there must be a Watchmaker.

Paley: You surprise me, Mr. Darwin. No one has given

me that answer before. I must now ask whether you see no plan at all?

Darwin: I will be quite candid since there is no longer any danger of being burnt as a heretic. It is my scientific opinion that man is the result of a purposeless and materialistic process that did not have him in mind. He was not planned.[4]

Paley: Do I understand you correctly, sir? You assert that to make a perfect and beautiful machine, it is not requisite to know how to make it?[5]

Darwin: Quite so.[6]

Paley: Sir, I confess that this is most astonishing. May I ask if you do not find your answer rather improbable?

Darwin: It is surely improbable that a perfect and beautiful machine could be made without foreknowledge of what was wanted; but only improbable, not impossible. It does not matter that it is *highly* improbable, for my system of natural selection is equal to the task. As R. A. Fisher has pointed out, it is a mechanism for generating improbabilities.[7]

Paley: But Mr. Darwin, how can a blind and automatic sifting process like selection, operating on a blind and undirected process like mutation, produce organs like the eye or the brain, with their almost incredible complexity and delicacy of adjustment? How can chance produce elaborate design? Are you not asking me to believe too much?[8]

Darwin: No, all this is not too much to believe, once one has grasped the way the system operates. I refer you again to Professor Fisher's pithy phrase—natural selection is a mechanism for generating an exceedingly high degree of improbability.[9]

Paley: I have now heard that pithy phrase twice and am beginning to see its full import. It seems to be a shield against all arguments based on improbability.

Darwin: Exactly. All the objections to a selectionist explanation of evolution that are based on the improbability of its results simply fall to the ground. In fact the shoe is now on the other foot. Improbability is to be *expected* as the result of natural selection; and we have the paradox that an exceedingly high apparent improbability can be taken as evidence for the high degree of its efficacy.[10]

Paley: But, my dear friend, I must protest that this is not at all what you say in your books. Only yesterday I examined *The Origin of Species* in order to see if it was compatible with ordinary human reason, which I take to be much the same thing as a feeling for probability. I was delighted to find that it was quite compatible. You said "probably" on the first page. True, the word did not appear on the last page, 490, but I found it twice on 488. And when I looked at the exact middle, pages 245 and 246, I found it on each page. You were appealing to reason throughout. There was nothing startling such as this pithy phrase that you ascribe to Fisher.

Darwin: Dr. Paley, you are entirely correct. When I wrote that book in 1859, I was a simple unsophisticated naturalist. But in the course of time, mathematicians like Fisher have shown me the force of arguments based on statistics, equations, and formulas.

Paley: I feel that there is something wrong, but I cannot at once formulate it. Would you agree, Mr. Darwin, to continue our conversation at a later date, when I have had leisure to brood over the pithy phrase?

Darwin: My dear sir, I will gladly resume at your convenience.

This is where the matter rests. So far as the evolutionists are concerned, they have vanquished Paley and destroyed all objections based upon the lack of probability. But if one probes further, there are indications that the Darwinians are not fully aware of what they have done. Their reasoning may not withstand careful scrutiny. I see three objections to it.

First, the pithy phrase is a debater's trick, having nothing to do with plants or animals. The train of argument is reducible to this: Nature is full of marvels; these could have been produced only by a Watchmaker or by natural selection; a Watchmaker has no standing in science, hence must be ruled out; ergo natural selection has produced all the marvels; ergo the greater the improbability, the greater the glory of natural selection. A vacuum is created by discarding all other suggestions; whereupon natural selection, with no examination of its credentials, is awarded the prize as the sole remaining candidate.

This is never stated clearly; one is only referred to Fisher's pithy phrase without any clue as to where to find it or any explanation of how Fisher worked it out.[11] But Sir Julian Huxley inadvertently reveals the flaw when he says that the pithy phrase "is a useful shorthand phrase to denote the real improbability of the results having been produced *in any other way* than by means of natural selection" (italics mine).[12] It may be correct that any other explanation is unlikely, but this does not make the natural selection explanation probable or justify declaring it to be the winner by default.[13]

Simpson's mind is operating in the same way when he says: "The origin of such an organ as the eye, for example, entirely at random seems almost infinitely improbable," then adds that there must have been "some additional factor or process" and that this must have been natural selection.[14] He creates a vacuum, offers natural selection as the only remaining possibility, and regards this as a proof that natural selection can do anything. It is unnecessary for him to show what natural selection actually can do. A logician would call this begging the question.

Second, in their ordinary transactions the evolutionists continue to speak of probability in a sensible way and to reason from it like other persons. Simpson, for example, speaks of science as "the critical weighing of probabilities in the light of objective evidence,"[15] although this is just what the pithy phrase excludes. Smith, in regard to the pattern of cleavage in different kinds of eggs, says: "This pattern is so characteristic that it seems *unlikely* to have arisen independently more than once; it is therefore concluded that the three phyla are descended from a common ancestor" (italics mine).[16] He does not see that this sort of reasoning falls to the ground if one takes the pithy phrase seriously, because nothing is then unlikely. Simpson follows suit in many places by arguing, quite properly if he had never heard of Fisher, that certain things are too improbable to be taken seriously.[17] Hardin censures Lysenko because he "rejects the theory of probability,"[18] although Hardin himself has left it pretty far behind. This inconsistency leads me to doubt that Huxley, Simpson, or even Fisher himself would really stand behind the pithy phrase if the matter were brought to a test.

Third, the effect of the pithy phrase is to seal off discussion and tests. It removes the question from the realm of science, since theories that are not testable are not scientific. Here I invoke Sir Karl R. Popper, the Austrian-born philosopher who, as professor of logic and scientific method at the London School of Economics, has been policing British scientists and historians for many years. Popper is not a biologist, but his interest in scientific method has led him to study the proper ways of formulating and refining theories. He has worked up a short list of earmarks by means of which theories can be evaluated as good or bad. I will quote only one: "A theory which is not refutable by any conceivable event is non-scientific. Irrefutability is not a virtue of a theory (as people often think) but a vice."[19]

As examples Popper points to the theories of Marx, Freud, and Adler, all of which had the following blemishes: Everything in the world was looked on as a verification; there were no experiments that could test them; and they could not be proved wrong. They might, of course, be correct, but there was no way of knowing whether they were right or wrong, and therefore they were not in the realm of science.

These blemishes are also present in Darwinism. Anyone familiar with the literature knows how everything is taken as a verification, being quickly fitted in without any critical reflection;* e.g., Sir Julian Huxley automatically interprets the coloring of grasshoppers as a camouflage without stopping to think that the animals give away their location by chirping.[20] Less well known is the fact that it is impossible to test large areas of Darwinism by experiments, since the crucial events either happened in the past or would need thousands of years before a conclusion could be reached.[21] If we are now told, as to the marvels, that the shoe is on the other foot and that extremely high improbability is an argument for rather than against natural selection, then there is no way to prove Darwinism wrong. But by the same token Darwin must join Marx, Freud, and Adler outside the scientific fold. It is my guess that the evolutionists would prefer to abandon the pithy phrase.

Let us now return to William Paley, Archdeacon of Carlisle. If he had thought about the pithy phrase for a couple of years, he would have been able to refute it. But if he had not asked the wrong question in the first place, the whole argument could have been avoided. He is not to be blamed for this, of course, since the words put into his mouth are only an historical reconstruction. Had he been present in person, he might have found a craftier approach.

Paley should never have said: "May I ask if you do not find your answer rather improbable?" The mention of probability led the discussion out of biology into logic and mathematics. It produced nothing but doubt and confusion.

Paley's proper question was: "May I ask you to give me a description of precisely how this could be accomplished in a

---

* Darwin (1871), 516, shows the amazing lengths to which this practice can go: "It can hardly be doubted that with most mammals the thickness of the hair on the back and its direction, is adapted to throw off the rain; even the transverse hairs on the fore-legs of a dog may serve for this end when he is coiled up asleep. Mr. Wallace, who has carefully studied the habits of the orang, remarks that the convergence of the hair towards the elbow on the arms of the orang may be explained as serving to throw off the rain, for this animal during rainy weather sits with its arms bent, and with the hands clasped round a branch or over its head."

specific case such as the eye or the brain?" This would have required a biological answer and would have left Darwin fumbling,[22] as we saw when discussing the marvels in the chapter on adaptation. The conversation would have terminated at that point, to Darwin's discomfiture.

This does not mean that Paley's position was correct. His arguments in favor of a Watchmaker are also outside the realm of science, and it would be easy to discomfit him if he wanted to pursue the case on a scientific basis. We are in a situation where neither party can sustain the burden of proof, and when this occurs it is the skeptic who comes out ahead. This has always been the result in the long battle of Design *v.* Chance; a few persons are wholly convinced one way or the other, but the mass of men in the middle, though impressed by the arguments on both sides, remain unpersuaded.

The vitalists and other persons who see a Watchmaker or the hand of God behind the marvels of nature should not be reckoned fools. They *feel* this presence, and the Darwinian arguments are not persuasive enough to overcome their feeling.

Unfortunately, however, these persons can easily be made to look like fools if the argument takes a wrong turn. Since they generally subscribe to certain theological doctrines as well as feeling the presence of a deity, their opponents can discomfit them by switching to theology and asking how a benevolent god could create some of the horrors in nature, and why an omniscient god would create ninety-nine species that died out for every one that survived. These questions torture the vitalists, but they are not essential to the problem and should be neither asked nor answered by modern biologists. Therefore I have omitted them here and I have a good precedent for so doing. When old Dr. Broom announced that it was clear to him that evolution was accomplished, not by selection or mutations, but by spiritual beings of various grades and various kinds of intelligence, Simpson had a great opportunity to twit him with such questions (though Broom would have given him a peppery response). But Simpson did not do this; he noted his dissent, made his reply, and avoided theology.[23] I frequently disagree with Simpson, but I admire his conduct in this case.

1. Hardin (1961), 58.
2. Simpson (1949), 267.
3. G. B. Shaw, Preface to *Back to Methuselah*.
4. Simpson (1949), 344–345.
5. From an anonymous writer in the *Athenaeum* for 1868, quoted in Hardin (1961), 260.
6. Hardin (1961), 260.
7. Hardin (1961), 260.
8. Huxley (1957), 40.
9. Huxley (1957), 40.
10. Huxley (1957), 43.
11. I am not the only one who cannot find it; see Grant (1963), 190.
12. Huxley (1957), 40; McAtee (1937), 47, comes very close to exposing the fallacy.
13. Huxley (1942), 413, asserts that one "must believe either in purposive creation or in adaptive evolution." Is it fair to prohibit the skeptic, the agnostic, and the common man from remaining in their traditional middle position?
14. Simpson (1964), 18–19; see also 188, 207–208, as well as Fisher (1954), 90–93.
15. Simpson (1964), 54.
16. Smith (1959), 265–266.
17. Simpson (1953), 96; (1964), 238, 267.
18. Hardin (1961), 210.
19. Popper (1963), 33–37.
20. Huxley (1942), 431–432; also 23, 149, 189, 192–193, and 212; compare Brooks (1957), 88, and Olson (1960), 536.
21. Deevey (1967), 639; Lack (1947), 118; Mayr (1961), 1503–1504; Stebbins (1950), 106–107.
22. Darwin (1859), 186–189; Fisher (1930), 41; Hardin (1961), 71–72, 224; Mayr (1960).
23. Broom (1933); Simpson (1949), 277, 326.

My basic definition of classical Darwinism in Chapter 1 included two corollaries stemming straight from Darwin himself. First, in the evolution of any structure of function, every intermediate stage must be of advantage to the species.[1] Second, natural selection tends only to make each organic being as perfect as, or slightly more perfect than, the other inhabitants of the same country with which it has to struggle for existence.[2] These were described as *logical* corollaries because they are derived from thinking about the implications of the theory, rather than from observation of nature. They are really predictions.

These corollaries are extremely important in evaluating classical Darwinism as a scientific achievement, and such an evaluation requires me again to take counsel with Popper, that connoisseur of scientific method. Popper shows with admirable brevity that the criterion of the scientific status of a theory is its falsifiability, or refutability, or testability.[3] He shows what this means by the examples of Einstein, astrology, Marx, Freud, and Adler. His perspicacity justifies a lengthy quotation:

> Einstein's theory of gravitation clearly satisfied the criterion of falsifiability. Even if our measuring instruments at the time did not allow us to pronounce upon the results with complete

assurance, there was clearly a possibility of refuting the theory.

Astrology did not pass the test. Astrologers were greatly impressed, and misled, by what they believed to be confirming evidence—so much so that they were quite unimpressed by any unfavorable evidence. Moreover, by making their interpretations and predictions sufficiently vague they were able to explain away anything that might have been a refutation of the theory had the theory and the prophecies been more precise. In order to escape falsification they destroyed the testability of their theory. It is a typical soothsayer's trick to predict things so vaguely that the predictions can hardly fail; that they become irrefutable.

The Marxist theory of history, in spite of the serious efforts of some of its founders and followers, ultimately adopted the soothsaying practice. In some of its earlier formulations . . . their predictions were testable, and were in fact falsified. Yet instead of accepting the refutations the followers of Marx reinterpreted both the theory and the evidence in order to make them agree. In this way they rescued the theory from refutation; but they did so at the price of adopting a device which made it irrefutable. They thus gave a "conventionalist twist" to the theory; and by this stratagem they destroyed its much advertised claim to scientific status.

The two psycho-analytic theories were in a different class. They were simply non-testable, irrefutable. There was no conceivable human behavior which could contradict them. This does not mean that Freud and Adler were not seeing certain things correctly; I personally do not doubt that much of what they say is of considerable importance, and may well play its part one day in a psychological science which is testable. But it does mean that those "clinical observations" which analysts naively believe confirm their theory cannot do this any more than the daily confirmations which astrologers find in their practice. And as for Freud's epic of the Ego, the Super-ego, and the Id, no substantially stronger claim to scientific status can be made for it than for Homer's collected stories from Olympus. These theories describe some facts, but in the manner of myths. They contain most interesting psychological suggestions, but not in a testable form.

Popper has performed a further service by converting his

generalizations into a checklist of seven propositions that are useful in assessing theories. I will quote five of them as pertinent to our problem.

1. It is easy to obtain confirmations, or verifications, for nearly every theory—if we look for confirmations.
2. Confirmations should count only if they are the result of *risky predictions;* that is to say, if, unenlightened by the theory in question, we should have expected an event which was incompatible with the theory—an event which would have refuted the theory.
3. Every "good" scientific theory is a prohibition; it forbids certain things to happen. The more a theory forbids, the better it is.
4. A theory which is not refutable by any conceivable event is non-scientific. Irrefutability is not a virtue of a theory (as people often think) but a vice.
7. Some genuinely testable theories, when found to be false, are still upheld by their admirers—for example by introducing *ad hoc* some auxiliary assumption, or by reinterpreting the theory *ad hoc* in such a way that it escapes refutation. Such a procedure is always possible, but it rescues the theory from refutation only at the price of destroying, or at least lowering, its scientific status. (I later described such a rescuing operation as a "conventionalist twist" or a "conventionalist stratagem.")

Let us now apply all this to classical Darwinism. Anyone familiar with the literature knows how everything is taken as a confirmation.[4] Also, there is no way to test large parts of the theory by experiment.[5] Therefore, if the theory is to escape the curse of irrefutability, it must be tested by its predictions or prohibitions. This brings us back to the corollaries. As predictions or prohibitions, they are among the few points at which the theory can be tested. Therefore it is important to see how they have held up under scrutiny.

First Corollary—Not Enough Mindlessness

Why must every intermediate step be advantageous to the species? Because Darwin conceived natural selection as a mindless process, as the impersonal operation of purely natural forces. If it is mindless, it cannot plan ahead; it cannot make sacrifices now to attain a distant goal, because it has no goals and no mind with which to conceive goals. Therefore every change must be justified by its own immediate advantages, not as leading to some desirable end.

This corollary has found its nemesis in the human eye. Hardin puts the problem neatly: "How then are we to account for the evolution of such a complicated organ as the eye? . . . If even the slightest thing is wrong—if the retina is missing, or the lens opaque, or the dimensions in error—the eye fails to form a recognizable image and is consequently useless. Since it must be either perfect, or perfectly useless, how could it have evolved by small, successive, Darwinian steps?" [6]

Hardin then offers an answer that I will quote in full:

> Were all other organisms blind, the animal which managed to evolve even a very poor eye would thereby have some advantage over the others. Oysters have such poor eyes—many tiny sensitive spots that can do no more than detect changes in the intensity of light. An oyster may not be able to enjoy television, but it can detect a passing shadow, react to it as if it were caused by an approaching predator, and—because it is sometimes right—live another day. By selecting examples from various places in the animal kingdom, we can assemble a nicely graded series of eyes, passing, by not too big steps, from the primitive eyes of oysters to the excellent (though not perfect) eyes of men and birds. Such a series, made up from contemporary species, is not supposed to be the actual historical series; but it shows us how evolution could have occurred. [7]

What are the weaknesses in this statement? I will point out two, although there may be more. 1) Doubtless one can collect samples from various species to build up a nicely graded series of eyes, but this has nothing whatever to do with the way the specific human eye was developed. Hardin admits this when he says that "such a series . . . is not supposed to be the actual historical series." Since it is the historical series we are asking for, he is giving us stones for bread. 2) Collecting a group of samples would actually show that nature had solved the problem in a number of different ways; but when we cannot explain even one way, the mystery only deepens when we see that nature has worked out several.

Hardin must have realized that his answer was inadequate, for he returned to the problem later in his book, saying: ". . . *That damned eye*—the human eye . . . which Darwin freely conceded to constitute a severe strain on his theory of evolution. Is so simple a principle as natural selection equal to explaining so complex a structure as the image-producing eye? Can the step-by-step process of Darwinian evolution carry adaptation so far? Competent opinion has wavered on this point." [8] Having thus marched up to the problem a second time, Hardin marched away from it with no answer at all. I read on for a number of pages expecting to see the waverings of competent opinion, but nothing appeared. I slowly realized that Hardin had changed the subject.*

The eye is the usual theme in this debate, but let me give a second example for the sake of variety. Certain sea slugs have appendages called papillae growing from their backs. In these papillae are groups of sting cells, usually of a long whiplike shape. In their undischarged condition, the stings are folded up so that the least touch will cause the coiled nettle-lash to fly out and sting any foreign body within reach. Since similar stings have been found in Coelenterates (little animals on which the sea slugs feed), it was supposed for a long time that the slugs were related to the Coelenterates. Recent research, however, has shown that there is no relationship and that the

---

* In a letter to Asa Gray dated 3 April 1860 Darwin said: "I remember well the time when the thought of the eye made me cold all over, but I have got over this stage of the complaint, and now small trifling particulars of structure often make me very uncomfortable. The sight of a feather in a peacock's tail, whenever I gaze at it, makes me sick!"

slugs have simply stolen the stings from the Coelenterates. They eat the Coelenterates, but somehow they keep from exploding the stings. They get the stings into their stomachs, then work them into narrow channels that have cilia or hairs in them. By means of the cilia they sweep the stings up the channels into pouches out on the papillae, and there the stings are all neatly arranged, right way up and still unexploded, in such a way that they can be discharged against an attacker.

I stumbled on this case while reading in quite a different field. Inquiring among biologists, I discovered that there are many similar cases, but they seldom appear in the standard literature. They are interesting, highly relevant, and well known, but they are the special stock in trade of the anti-Darwinists.[9] These heretics delight in flaunting such cases in the face of the evolutionists and demanding explanations on the usual step-by-step utilitarian lines. Since nobody really pretends to know how such things came about, the usual response is silence. Not one of my four paperbacks mentions a case of this type.

Note, however, that these cases have been posed only as normal problems, with no emphasis on the corollary. If we now add the requirement that every intermediate step must be advantageous, the already insoluble problem becomes even more so. Thus it seems fair to say that the first corollary has been disproved.

Second Corollary—Too Much Perfection

Darwin formulated this himself in the first edition of *The Origin of Species*: Natural selection tends only to make each being as perfect as, or slightly more perfect than, the other inhabitants of the same area. Eiseley reports that in 1869, after only ten years, it was brushed aside by no less a person than Alfred Russel Wallace, co-inventor with Darwin of the doctrine of natural selection.[10] Perceiving that the gap between the brain of the ape and that of the lowest savage was too big, Wallace announced a heresy: "An instrument has been developed in advance of the needs of its possessor." He challenged the whole Darwinian position by insisting that artistic, mathematical, and musical abilities could not be explained on the basis of natural selection and the struggle for existence. Something else, he contended, some unknown spiritual element, must have been at work in the elaboration of the human brain. He added, perhaps with a touch of malice toward his colleagues: "Natural Selection could only have endowed the savage with a brain a little superior to that of the ape, whereas he actually possesses one very little inferior to that of the average member of our learned societies."

Darwin realized that this was dangerous. He wrote to Wallace: "I hope you have not murdered too completely your own and my child." If I read Eiseley correctly, Wallace never found an answer to this difficulty and remained a skeptical Darwinist until his death in 1913. Nor does Eiseley suggest that any answer is now available.[11] The unavoidable conclusion is that the second corollary has also been disproved.

Thus both corollaries have been tried and found wanting. The predictions have been falsified. This may be why modern evolutionists admit that Darwinism has no predictive power. Mayr, for example, says: "The theory of natural selection can describe and explain phenomena with considerable precision, but it cannot make reliable predictions."[12] Deevey puts it more tersely: ". . . evolutionists are still historians, not prophets."[13]

Let us now revert to Professor Popper's Proposition No. 7: "Some genuinely testable theories, when found to be false, are still upheld by their admirers." Has this happened to classical Darwinism? Yes, and the process began with Darwin himself. Eiseley describes it thus: "He did not, however, supply a valid answer to Wallace's queries. Outside of murmuring about the inherited effects of habit—a contention without scientific validity today—Darwin clung to his original position. Slowly Wallace's challenge was forgotten and a great complacency settled down upon the scientific world." [14]

This comfortable ignoring of the results is one way to get rid of the falsification, but we should also look for a "conventionalist twist" such as Popper mentions. Mayr supplies something of this kind. He knows that in physics the power to predict is closely tied to the power to explain, but he denies that this is true in biology. He tries to dissociate the two entirely, implying that Darwinism can be a perfect explanation despite being no good at all at predicting. His words are very bold: ". . . one of the most important contributions to philosophy made by the evolutionary theory is that it has demonstrated the independence of explanation and prediction." [15]

I suspect that many philosophers would be startled by this statement,[16] but the present case is not a good one for testing it. We agree that Darwinism cannot predict, but we must remind Professor Mayr that it also cannot explain; hence we do not have explanation without prediction. If the reader wonders how I can dare to make such an accusation, let him recall that Dobzhansky cannot explain why the more than six hundred known species of *Drosophila* all have three orbital bristles on either side of their heads;[17] that Simpson cannot explain why his favorite squirrels have ear tufts[18] or why average stature in the United States has increased since 1900;[19] and that no one has ever responded to Goldschmidt's demand for a Darwinian explanation of seventeen specific cases.[20] Listen also to what one eminent theorist says, very benevolently, about Darwinian explanations:

> I doubt that there is a scientist who would question the ultimate causality of all biological phenomena—that is, that a causal explanation can be given for past biological events. Yet such an

explanation will often have to be so unspecific and so purely formal that its explanatory value can certainly be challenged. In dealing with a complex system, an explanation can hardly be considered very illuminating that states: "Phenomenon A is caused by a complex set of interacting factors, one of which is B." Yet often this is about all one can say.

Who is this eminent theorist? It is Professor Mayr,[21] who just implied that Darwinism was good at explaining, even if weak at predicting.

Mayr may have saved classical Darwinism from refutation, but I suspect that the cost was too high. Was it not, in Popper's words, at the price of destroying, or at least lowering, its scientific status?

1. Darwin (1859), 199–201; Hardin (1961), 71.
2. Darwin (1859), 201; Eiseley (1961), 310.
3. Popper (1963), 36–38; Simpson (1964), 4, 256, and (1969), 21, seems to agree.
4. Huxley (1942), 431–432; Brooks (1957), 88; Olson (1960), 536. This is carried to an amazing extreme in Whittaker and Feeny (1971), 759. Speaking of plants that poison themselves, they piously expect to find that this somehow benefits the plants: "Self-toxicity is an evolutionary paradox. One supposes that some selective advantage from production of toxic compounds outweighs the disadvantage of self-inhibition."
5. Deevey (1967), 639; Lack (1947), 118; Mayr (1961), 1503–1504; Stebbins (1950), 106–107. One of the best remarks on this point was made by von Bertalanffy as quoted in Olson (1960), 530: "A lover of paradox could say that the main objection to selection theory is that it cannot be disproved."
6. Hardin (1961), 71.
7. Hardin (1961), 71–72.
8. Hardin (1961), 224.
9. Olson (1960), 523–524, speaks with some sympathy of this "vocal, but little heard, minority" and also of a larger group who tend to disagree with current thought but, for various reasons, say and write little.
10. Eiseley (1961), 310–314, and (1958), 84–85.
11. Eiseley (1958), 85.
12. Mayr (1961), 1504.
13. Deevey (1967), 639; see also Simpson (1969), 47–48. Manser (1965) supports Deevey's position by demonstrating that Darwinism can neither predict nor explain.
14. Eiseley (1958), 84–85. Darwin (1871), 432, actually went so far as to say: "I cannot, therefore, understand how it is that Mr. Wallace maintains that 'natural selection could only have endowed the savage with a brain a little superior to that of an ape'."
15. Mayr (1961), 1504.
16. See, for example, Barker (1969) and Lee (1969).

17. Dobzhansky (1956), 339–340.
18. Simpson (1953), 170.
19. Simpson (1964), 276.
20. Goldschmidt (1940), 6–7.
21. Mayr (1961), 1503. Professor Mayr's benevolence comes out clearly when one compares his mild condemnation of such explanations with the crisp censure of Fischer (1970), 175: "*The fallacy of indiscriminate pluralism* is the converse of the reductive fallacy. It appears in causal explanations where the number of causal components is not defined, or their relative weight is not determined, or commonly both. The resultant explanation, for all its apparent sophistication and thoroughness, is literal nonsense."

Darwin, thinking of evolution as the accumulation of myriads of small changes, needed a great deal of time to bring the plants and animals to their present complexity and diversity. This led him to support vigorously the geological opinions of his slightly older contemporary, Sir Charles Lyell (1797–1875). Lyell's view, known as "uniformitarianism," was that the visible features of the earth had been produced by the action, at more or less the present scale and tempo, of the agencies we still see at work—wind, weather, water, ice, volcanoes, and earthquakes. Since the performance was necessarily slow, Lyell postulated much longer spans of time than were commonly considered possible when he began to publish, around 1830 to 1840. Darwin, perceiving that this was also to his benefit, eagerly adopted uniformitarianism. He even went beyond Lyell by stressing how slowly everything worked, whereas Lyell liked to show that nature could be brisk at times.

The inflationary tendencies of Lyell and Darwin were rudely checked, only a few years after the publication of *The Origin of Species,* by a third giant of English science, the physicist William Thomson, Lord Kelvin (1824–1907). Applying physics to the study of the earth's age, Lord Kelvin calculated such items as mass, temperature, and heat loss, and came out with shattering results. At first he saw thirty million years as about the maximum. Then he reduced the figure to twenty million, or even fifteen.[1]

Thomas Henry Huxley, Darwin's pugnacious partner, was inclined to meet this attack by saying merely that the processes of evolution must have been correspondingly speedier. Darwin himself, greatly to his credit, did not try to save himself by such sophistry, although he referred to Lord Kelvin as an "odious specter" and feared that Darwinism was finished if Kelvin was correct: "I am greatly troubled at the short duration of the world according to Lord Kelvin, for I require for my theoretical views a very long period."

As every reader knows, the Kelvin calculations proved to be unsound.[2] The physicists now allow enormous stretches of quiet time for the history of the earth. The biologists are happy on this score, but perhaps unwisely so. There are signs, no bigger than a man's hand, that the past of the earth was not always leisurely and tranquil. The physicists may be wrong again, as to both duration and tranquillity.

The subject of chronology is extremely complicated, especially because of the still incomplete work with new dating methods such as carbon-14 and argon. I will not attempt to discuss either theory or practice, being content only to point out a disturbing line of evidence reported by Eiseley:

> But suppose, just for a moment, that this period of the great ice-advances did not last a million years—suppose our geological estimates are mistaken. Suppose that this period we have been estimating at one million years should instead have lasted, say, a third of that time. In that case, what are we to think of the story of man? Into what foreshortened and cramped circumstances is the human drama to be reduced? Such an episode, it is obvious, would involve a complete reëxamination of our thinking upon the subject of human evolution. In 1956 Dr. Cesare Emiliani of the University of Chicago introduced just this startling factor into the dating of the Ice Age. He did it by

the application of a new dating process developed in the field of atomic physics.[3]

The process measured the amount of oxygen-18 in oyster shells, ascertaining from this the temperature of the water in which the oysters once lived. Cores of undisturbed sediment were brought up from the ocean floor. Careful study showed changes in temperature from layer to layer in the cores, enabling Dr. Emiliani to chart the climates of the past and trace the ups and downs of the Ice Age. The results were disturbing because they cut the duration to a third or less of what had been assumed to be the proper figure.

I do not assert that Dr. Emiliani was correct. I only want to show that geological and biological projections into the past (like most extrapolations) have a precarious base and that a shift in chronology would shake the foundations on which Lyell and Darwin constructed their theories.

Dr. Emiliani is probably not popular among the natural scientists, but now we must move from chronology to castastrophes and discuss a man whom they have handled very roughly. This is Dr. Immanuel Velikovsky of Princeton, New Jersey, author of *Worlds in Collision* (Doubleday, 1950).[4] This book stirred a number of astronomers and physicists to such denunciations that I would hesitate to mention Velikovsky if certain other scientists had not later come to his defense. The *American Behavioral Scientist* for September of 1963 reviewed the attacks on Velikovsky and censured the intemperate actions of the natural scientists.[5] There is a bold speculative cast to Velikovsky's work, but when he is cited as a reporter rather than as an original authority I believe that his name can be mentioned without having to make a long defense of his position.[6]

Velikovsky's opponents pointed out that he was talking about events that qualified as catastrophes, transcending anything that is now going on in scale and violence. They declared that this put Velikovsky out of court because the uniformitarian doctrine provided no room for such events. Velikovsky, who had practiced medicine and was rather innocent as to Anglo-Saxon geological theory, was surprised at this reaction and at the violent feelings he had aroused. His response was admirable; without extensive public recrimination, he disappeared into the library for several years and compiled a book called *Earth in Upheaval* (Doubleday, 1955). Here he marshals the original field reports on a large number of phenomena that point inexorably to catastrophes and (as a byproduct, since he was looking for events rather than dates) to fairly recent dates for the catastrophes. The impact of the details and of the number of phenomena (close to forty) is shattering. I hold no brief for Velikovsky's theories, but I am indebted to him for collecting material that had never been assembled in one place before.

The topics in the book are discussed on the basis of reports by orthodox and reputable scientists, with Velikovsky merely acting as master of ceremonies. I will epitomize the material on six themes that were especially interesting to me.

Lava Beds of the Columbia Plateau

Something like 200,000 square miles in Idaho and eastern Washington and Oregon are covered with lava, which in many places is 5000 feet or more in depth. All the volcanoes in the world, working at their present paltry scale and tempo through any period of time, could never produce such quantities of lava; hence this is a direct challenge to the uniformitarian theory. To make matters worse, much of the lava seems to be fresh, and a figurine of baked clay was found at a depth of 320 feet. If men were present and making figurines before the eruptions ceased, the eruptions must have been very recent.

I found this report fully confirmed by Ruth Moore, an able popularizer in the earth sciences.[7] Miss Moore actually gives far more astonishing details on the lava than Velikovsky does, but she never mentions the figurine or the freshness. She is an ardent admirer of Lyell and maintains a conviction in the uniformitarian theory that allows her to say: ". . . nothing has ever indicated that Lyell was wrong about the general uniformity of the earth's behavior." [8]

The Harras of Arabia

Many signs and traditions indicate that Arabia was once a green and pleasant land, although it is now quite the opposite. The change was so recent, drastic, and complete as to be puzzling, and when one is aware of the curious collections of stones known as the harras one can no longer make the usual easy suggestion that the climate must have changed. Velikovsky's brief description will suffice to show that something strange has happened in Arabia.

> Twenty-eight fields of burned and broken stones, called harras, are found in Arabia, mostly in the western half of the great desert. Some single fields are one hundred miles in diameter and occupy an area of six or seven thousand square miles, stone lying close to stone, so densely packed that passage through the field is almost impossible. The stones are sharp-edged and scorched black. No volcanic eruption could have cast scorched stones over fields as large as the harras; neither would the stones from volcanos have been so evenly spread. The absence, in most cases, of lava—the stones lie free—also speaks against a volcanic origin of the stones. . . . Despite alternate exposure to the thermal action of the hot desert sun and the cool desert night, the sharp edges of the stones have been preserved, which shows that they fell in a not too distant period of time.

Youthfulness of Mountain Chains

I knew from my college course in geology that there were old mountains and young mountains, but I was not prepared for the extreme youthfulness that seems to exist. The Andes and the Himalayas contain traces of human habitation at levels (16,000–18,000 feet) that would not now be feasible for settlement. The Alps show the same thing on a smaller scale. Apparently all these chains have risen extensively since men moved in, and much of the upthrusting has occurred in the short period since the retreat of the glaciers. It is impossible to express this precisely in years, but the span of time is almost infinitesimal when compared to the figures commonly used by geologists. Needless to say, the upthrusting was not a quiet everyday event.

Checking a couple of current college textbooks used by my children, I found that practically nothing was said about mountain-building and that the subject seems to baffle the scholars. Ruth Moore confirmed this impression: "The bewildering old question of what has elevated the mountains and the continents still has not been answered." [9]

Klimasturz

There is evidence that in the past there were changes of climate so violent and sudden as to merit the name of *Klimasturz*, which can be translated as a plunge, tumble, crash, or collapse of the climate. Velikovsky summarizes the work of H. Gams and R. Nordhagen on lakes and fens in Germany and Switzerland, showing that such a change occurred about 800 B.C.

They undertook a close examination of the pollen content of peat-bogs. Since the pollen of each species of tree is characteristic, it is possible to detect by analysis what kinds of forests grew in various periods of the past, and consequently the then prevailing climate. The pollen disclosed a "radical change of life conditions, not a slow building of fens." Man and animal suddenly disappeared from the scene, although at the time the area was already rather thickly populated. Oak was replaced by fir, and fir descended from the heights on which it had grown, leaving them barren.

The Alpine passes were much traveled during the Bronze Age: many bronze objects from before 700 B.C. were found in numerous places, especially on St. Bernard. Also mines were worked in the Alps in the Bronze Age. With the advent of the *Klimasturz* the mines were suddenly abandoned, and the passes were not traveled any longer, as though life in the Alps had been extinguished.

Wandering of the Poles

Professor S. K. Runcorn in England has studied the movements of the Poles. He is Velikovsky's principal authority for three propositions that are a little startling to the layman. 1) The North and South magnetic Poles have reversed their fields many times. 2) The magnetic Poles have wandered. The North Pole was at one time on the coast of California, then moved across the entire Pacific Ocean and went up through Siberia to its present location. 3) The geographical poles moved in the same way, so that the axis of rotation changed profoundly and caused the planet to "roll about." The implications of such gigantic movements of the poles and of such language as "roll about" are great enough to cause doubt as to the uniformity and steadiness of the earth's behavior in the past.

Velikovsky and Ruth Moore quote the same passage from Runcorn: ". . . the earth's axis of rotation has changed also. In other words, the planet has rolled about, changing the location of its geographical poles." [10] It is amusing, however, to see how widely their reports vary in other respects. Velikovsky adds material from Swiss scientists showing that the last reversal of the magnetic fields took place about 800 B.C., and he includes a passage where Runcorn says "the field would *suddenly* break up and reform with opposite polarity" (italics mine). Ruth Moore mentions no date later than two hundred million years ago and emphasizes that the movements were slow and gradual. She says nothing about *suddenly*, although she does admit that a shift could have occurred "in a relatively short time." How strong the influence of preconceptions can be!

The Frozen Mammoths

In various parts of Siberia, in the frozen gravel of the permafrost, the bodies of mammoths have been discovered in a remarkably perfect state of preservation. Not only were the flesh, skin, and hair in good condition, but even the eyeballs were intact. Food was found in the stomachs, and sometimes even in the mouths. As several writers have pointed out, the beasts must have been quick-frozen to achieve such preservation. But how could a mammoth, quietly pasturing on buttercups and other forage, be killed and quick-frozen before he could swallow?

Velikovsky suggests, like many before him, that the explanation could only be an event approaching the level of a catastrophe and therefore out of harmony with the uniformitarian theory. Furthermore, this could not have been a local event, because the frozen carcasses were found over an area of several thousand miles; and it could not have been in the remote past, because mammoths were known to the Ice Age painters and these carcasses are agreed to be only about ten to thirty thousand years old.

Lyell knew about the mammoths and saw that they endangered his theory. He tried to explain them away, suggesting they were caught in a cold snap while swimming;[11] which does not tally with the facts. Darwin also knew the story, and confessed that he saw no solution to it.[12] There have been a number of articles on the subject, with one group of authors crying Behold! and another group crying Pooh pooh! [13] In the latest entry in my files, Simpson does not discuss the mammoths directly, but criticizes the errors of two recent authors and then blasts the doctrine of uniformitarianism in a way that must have been highly agreeable to Velikovsky:

> Farrand expresses a common, probably the usual, modern understanding of uniformitarianism as "the geologist's concept

that processes that acted on the earth in the past are the *same processes* that are operating today, on the *same scale* and at approximately the *same rates*" [italics mine]. But the principle is also flatly contradicted by geological history. Some processes (those of vulcanism or glaciation, for example) have evidently acted in the past with scales and rates that cannot by any stretch be called "the same" or even "approximately the same" as those of today. Some past processes (such as those of Alpine nappe formation) are apparently not acting today, at least not in the form in which they did act. There are innumerable exceptions that disprove the rule.[14]

The reader should peruse Velikovsky himself so as to get the cumulative effect of his evidence, and he should also look at some of the original material, but he will then have to make up his own mind as to what is correct and who is sound. The wealth of specific cases pointing toward catastrophes makes it impossible for me to accept the uniformitarian theory, but I have the impression that in academic circles Ruth Moore's unquestioning faith is much commoner than Simpson's skepticism.[15] Catastrophes have been taboo for a century among the orthodox. Not one of my paperbacks mentions the problem, and the big treatises bring it up only when speculating on possible causes of extinction.

But a change may be impending. *Newsweek* for 13 December 1963 reported that ". . . many geologists at the recent meeting of the American Geological Society were advising the rehabilitation of catastrophism." Dr. Norman Newell of the American Museum of Natural History is said to have stated to a *Newsweek* writer: "Since the end of World War II, when a new generation moved in, we have gathered more data and we have begun to realize that there were many catastrophic events in the past." Such language must have made Darwin and Lyell turn in their graves. It should also be noted that at least one reputable archaeologist is willing to speak of drastic changes in very recent times and very familiar areas.[16]

1. Eiseley (1961), 233–244, is my source for the Kelvin material.

2. Eiseley (1961), 234.

3. Eiseley (1958), Chapter 8.

4. Dr. Velikovsky resides in the town of Princeton, but is not affiliated with the little college there.

5. This is now reprinted as a book: De Grazia (1966).

6. Some might not allow even this: Deevey (1967), 633.

7. Moore (1956), Chapter 17.

8. Moore (1956), 345.

9. Moore (1956), 343.

10. Velikovsky (1955), 145; Moore (1956), 325.

11. Lyell (1953), Chapter 6.

12. Whitley (1910), 56.

13. The literature deserves a bibliography of its own. I will mention only Sanderson (1960) and Farrand (1961) since they well represent the two positions.

14. Simpson (1964), 132. The phrase "italics mine" is Simpson's. Ghiselin (1969), 14, makes some very puzzling remarks: "The principle of uniformitarianism is thus an instrument of investigation, not a scientific theory. . . . Uniformitarianism is not an empirical proposition at all, and is therefore neither true nor false."

15. Lyell himself may have entertained some doubts; certainly he gave at least one young acquaintance reason to wonder whether "the great Uniformitarian was strict or lax in his uniformitarian creed": see Chapter 15 of *The Education of Henry Adams*.

16. Carpenter, Rhys (1968), *Discontinuity in Greek Civilization*, Norton.

I have several times remarked that classical Darwinism has no explanatory power when confronted with specific cases. The professionals concede this,[1] but the reader will not get the full import of such a defect unless a clinching example is presented to him. The best clincher is extinction.

For every species now in existence, roughly ninety-nine have become extinct.[2] The question of *why* they have become extinct is of enormous importance to evolutionists. It has been studied by many men, but a convincing answer has not been found.[3] It remains unclear why any given species has disappeared.[4]

The discussion of survival of the fittest showed that the phrase led to circular reasoning; you survive because you are fit and you are fit because you survive. Discussion of extinction is beset by a similar danger. It is all too easy to say that a species becomes extinct because it fails to adapt, while establishing its failure to adapt only by its becoming extinct: in other words, you die because you are unfit and you are unfit because you die.[5] The temptation to reason thus is almost irresistible; hence we must bring out its absurdity by an analogy.

Our system of vital statistics requires a certificate for every death. The attending physician must state the cause of death, and for this purpose there is a standard list of causes, such as cancer, tuberculosis, asphyxiation, burns, or lead poisoning. One of these must be chosen. The physician may not say "Don't know," because when he does not know there must be an autopsy under the auspices of the coroner. Nor may the physician say "He stopped breathing," because this is true in every case and explains nothing. For the same reason he may not dress up his ignorance in elegant and meaningless phrases such as "Not viable" or "Lost adaptation."

Thus in current deaths there are three elements: the corpse, the categories, and the coroner. In extinction we have parallel elements in the vanished species, the possible causes, and the evolutionists. The evolutionists, however, are unable to discharge their function. As coroners they are useless. They offer explanations such as failure to adapt, overspecialization, loss of survival value, or inability to reproduce their kind, but when carefully examined these all turn out to be fancy ways of saying "He stopped breathing." They are correct, but not explanatory.

Mayr furnishes a striking example of such reasoning. He asserts that ". . . ultimately their extinction is due to an inability of their genotype to respond to new selection pressures." [6] This sounds impressive and is entirely correct, but it is meaningless because the same could be said of every extinct species and of every dead person, including Julius Caesar and Abraham Lincoln. Mayr must have realized that it was hollow, because five lines later he added: "The actual cause of the extinction of any fossil species will presumably always remain uncertain."

It is easy to make up lists of possible causes of extinction. Osborn did this very thoroughly more than sixty years ago, putting in disease, change of climate, competition, parasites, poisons, and several other possibilities.[7] The difficulty is in connecting the cause with the corpse and here we fail. Eiseley is very frank: "There is a simple reason for this. Theories are many, but most unprovable. Or pertinent objections to their general usefulness can be raised, even if we grant their applicability in particular cases. It is this situation that causes the biologist to despair as he surveys the extinction of so many species and genera. . . ." [8]

The inability to explain extinction causes awkward arguments, such as that on "hypertely," the idea that natural selection has bungled by carrying change too far. The common examples are the Irish elk and the mammoth, which developed such enormous and apparently useless antlers and tusks as to give the impression that they were unfit under any definition. Haldane, Waddington, and Huxley were baffled by these cases, confessing that they could not explain them "on any theory of evolution whatever." [9] Simpson, however, would not admit that natural selection had bungled. He went to work on these and similar cases to show that our judgment was too fallible to justify any firm verdict as to why the animals had become extinct. [10] His best argument was that we cannot *know* that the structures were disadvantageous when the species endured in large numbers for long ages.

But if we cannot tell in cases like these, we cannot tell at all. Simpson accepts this dismal conclusion unflinchingly: "Particular cases of extinction, other than those evidently due to competition, are usually hard or impossible to explain in detail. It is not that there is any serious doubt about the general cause, but that possible particular causes present an embarrassment of riches and many of them can leave no clear trace in the record of earth and life history. We do not know just why horses became extinct in the Americas around the end of the Pleistocene, not because the event is inexplicable but because there is no conclusive way of choosing among possible explanations." [11] Few coroners would say that they simply could not choose among all the possible explanations.

Simpson goes even one step further. Admitting we cannot tell why vanished species died out, he adds that we cannot tell why some living species do *not* die out. He shows that many flourishing species are so odd as to have no right to live: ". . . animals about as bizarre as any that ever became extinct are alive today and doing well: elephants are as queer as mammoths; living whales are far bulkier than any dinosaur; the spiral 'unicorn' tooth of the narwhal has no equal for strangeness among past animals; it would be hard to imagine anything more fantastic than some insects such as a dynastes beetle or some of the mantids." [12] I admire such a complete confession of ignorance, but it disqualifies Simpson as an explainer.

One more point must be mentioned. There are in nature certain forms that have existed unchanged through enormous stretches of time; e.g., the platypus, the little brachiopod *Lingula,* the oyster, the opossum, the ginkgo tree, the Australian lungfish, and the recently discovered fish called *Latimeria.* These are known as "living fossils" or "persistent types." They puzzle and annoy the evolutionists, who feel obligated to explain why, in a world of change, these forms continue in their old placid way without either changing or becoming extinct. In hundreds of millions of years there must have been changes in climate, changes in the environment, new enemies, new parasites, new diseases. Yet these creatures, without showing any special virtues or abilities, continue unchanged.

Sewall Wright, the mathematical biologist, has brooded on this: "The long-continued cessation of evolution in some forms is another problem. The apparently obvious explanation that mutation rate is exceptionally low is probably not the primary one." [13] He barely misses saying that the forms have not changed because they have not mutated. The language looks impressive, but I treasure it as a close approximation to *He stopped breathing*. Other authors frankly confess, without thus beating about the bush, that they have no explanation.[14] The Darwinists are stumped. They cannot explain extinction or survival, although these phenomena are the essence of evolution.

1. Mayr (1961), 1503–1505; Deevey (1967), 635; Olson (1960), 530.
2. De Beer (1966), 27.
3. Osborn (1906); Simpson (1949), Chapter 13, and (1953), Chapter 9; Eiseley (1943) and (1946).
4. Simpson (1949), 208.
5. Simpson (1949), 205; (1953), 294–299; (1964), 162.
6. Mayr (1963), 620.
7. Osborn (1906).
8. Eiseley (1946), 54.
9. Simpson (1953), 282 ff.
10. Simpson (1953), 286. Olson (1960), 538, lists many other fossil forms and clearly perceives their puzzling implications.
11. Simpson (1949), 202.
12. Simpson (1949), 202.
13. Wright (1965), 78, 97. It is surprising that Wright should speak thus in 1965, since the fallacy of this kind of "explanation" had been exposed at least twenty-three years before that date. Mayr (1942), 217, says: "The reasons why certain species and genera are stable, while others that live under similar conditions change rapidly, are still shrouded in mystery. It is a very unsatisfactory explanation to say that this is due to differences in the mutation rates, because we may ask immediately what causes these different rates in the different species."
14. Hardin (1961), 56; Simpson (1949), 101, 192–195; Stebbins (1950), 518.

In browsing through my biology books I have been startled by a number of things, but my biggest surprise was the discovery that religion is still a crucial matter. It is not always mentioned, but it exerts a large influence.

To begin with, the fundamentalists have not given up the ghost. On 11 November 1964 the United Press reported from Texas that several religious leaders had objected to the State Board of Education's approving five high-school biology textbooks containing Darwin's theory of the evolution of man. This objection was overruled, but on 13 November 1969 the California State Board of Education, responding to similar pressure, decided that future textbooks should present Darwinism as only one of several theories. At first blush this sounds like William Jennings Bryan at the Scopes trial in 1925, but there is a difference. This time the protest was not directed at the bare idea of evolution; it was on the more sophisticated ground that Darwinism, as an explanation of evolution, was taught as an accepted fact rather than as a debatable theory. The protesters could actually have cited the authority of Mayr, since he says much the same thing: "The basic theory is in many instances hardly more than a postulate and its application raises numerous questions in almost every concrete case." [1] (A postulate, by the way, is defined by Webster as "a position or supposition assumed without proof.")

Many strong believers favor a view that is often called "creationism," since divine creation is retained even if a literal interpretation of Genesis is no longer insisted upon. A specimen of their work is *Evolution and Christian Thought Today* (Eerdmans, 1959), a hardcover book containing eleven essays on different aspects of evolution. The authors are professional scientists (though not in first-class universities) as well as earnest Christians, and are by no means ignorant of the facts of life. They accept the broad and easy aspect of evolution, but reject Darwin's explanation as to how and why. They put their fingers on the various gaps that are commonly filled by assumptions or extrapolations, and they assert that the whole process could be the result of design just as well as of chance. Being careful writers and seeing the problems pretty clearly, they will never be easy marks like Wilberforce and Bryan.

Even Jehovah's Witnesses have learned a good deal of biology. The issue of *Awake* for 22 April 1967 was pressed upon me one day and I was amazed to find that it contained some shrewd criticism of Darwinism. The basic view was close to fundamentalism, but the anonymous authors were able to quote judiciously from Sir Arthur Keith; Sir Julian Huxley; Eiseley; J. H. Woodger; Conklin and Bonner of Princeton; Romer, Simpson, and Mayr of Harvard; Dobzhansky; Waddington; H. J. Muller; Goldschmidt; Le Gros Clark; Willard Libby; L. S. B. Leakey; and others. Thus it is no longer correct for Simpson to say: ". . . those who do not believe in it [evolution] are, almost to a man, obviously ignorant of the scientific evidence." [2] *

---

* Simpson talks like this only when he is annoyed. When he is calm, he speaks kindly of students disillusioned with Darwinism: "It would certainly be a mistake merely to dismiss these views with a smile or to ridicule them. Their proponents were (and are) profound and able students." Simpson (1964), 199.

The advocates of religion operate in the open, with their hearts on their sleeves. Their opponents seldom announce themselves as such, but their leanings are revealed from time to time. Rather to the surprise of some of his colleagues, Sir Julian Huxley declared in Chicago that he was an atheist and that Darwin's real achievement was to remove the whole idea of God as the creator of organisms from the sphere of rational discussion.[3] A contempt for Christianity sometimes bursts out of Simpson, as when he refers to "the higher superstitions celebrated weekly in every hamlet of the United States," [4] but in general he is restrained. This is only fair when one considers how useful the Christians are to him as whipping boys; frequently, when he is unable to explain something, he employs the stratagem of showing how much weaker the Christians (vitalists or finalists) would be.[5]

The reader must forgive the continual mention of Simpson. I quote him because he is among the most thorough and extensive writers on the theory, and because Deevey recommended him to me as having almost all the answers.[6] Therefore it is important to note how he excludes design as a matter of scientific principle and method: ". . . the progress of knowledge rigidly requires that no non-physical postulate ever be admitted in connection with the study of physical phenomena. We do not know what is and what is not explicable in physical terms, and the researcher who is seeking explanations must seek physical explanations only. . . ." [7] If a Watchmaker is thus carefully excluded at the beginning, we need not be surprised if no Watchmaker appears at the end. The dice have been loaded against him.

The determination to exclude Christianity plays a part in the arguments, but it is only a reflection of a far more significant fact: *Darwinism itself has become a religion.* This is sometimes openly admitted, as in the following words of Edwin G. Conklin (1863–1952), late professor of biology at

Princeton: "The concept of organic evolution is very highly prized by biologists, for many of whom it is an object of genuinely religious devotion, because they regard it as a supreme integrative principle. This is probably the reason why severe methodological criticism employed in other departments of biology has not yet been brought to bear on evolutionary speculation." [8]

One quickly notices a religious fervor in the literature,[9] but most biologists do not share Conklin's awareness of this. Thus T. H. Huxley once made the perspicuous remark that the new truths of science begin as heresy, advance to orthodoxy, and end up as superstition;[10] but I have found no one who asked which of these stages Darwinism is now in. In order to document my accusation I have compiled a list of five traits that seem to me to be earmarks of a religious attitude among the evolutionists. There are no clear criteria in such matters, hence these are only suggestions to the reader.

All Who Are Not with Me Are Against Me

Nobody uses language of this clarity, but it is made plain that all good men have a duty to inquire and to take sides, instead of looking on with the unreasoning wonder of a child.[11] Thus Simpson rejects the suggestion that "one could gather more facts and suspend judgement as to what meaning they might eventually have." [12] But the religious tone and attitude are even clearer in a passage used by Simpson as a motto at the head of a chapter: ". . . to abandon the scientific problem as insoluble . . . there can be no greater impiety than that. It is surrendering our birthright—not for a mess of pottage, it is true, but for peace of mind. Therefore man is true to himself when he presses home the question: How has this marvelous system of Animate Nature come to be as it is?" [13] This implies that a great many people, probably including all skeptics, are not true to themselves.

Reproof of the Fainthearted

When a colleague is unable to go along with Simpson's views, he is looked upon as no longer wholly reasonable. He is seen as succumbing to "despair or hope, an emotion even more blinding than despair";[14] or as a victim of defeatism or escapism.[15] This comes out best in a passage where Simpson laments the errors of Henry Fairfield Osborn, Teilhard de Chardin, and others: "In some cases these theories were clearly born of despair and faintness in the search, an emotional state with which we must sympathize but which we should surely seek to avoid in ourselves." [16] I call attention to *faintness in the search* as having the tone of the true zealot. Compare it with the admirable restraint recommended by the first Huxley:

> But you must recollect that when I say I think it is either Mr. Darwin's hypothesis or nothing; that either we must take his view, or look upon the whole of organic nature as an enigma, the meaning of which is wholly hidden from us: you must understand that I mean that I accept it provisionally, in exactly the same way as I accept any other hypothesis. Men of science do not pledge themselves to creeds; they are bound by articles of no sort; there is not a single belief that it is not a bounden duty with them to hold with a light hand and to part with cheerfully, the moment it is really proved to be contrary to any fact, great or small.[17]

Missionary Zeal

Simpson is convinced that evolution should be taught in every high school. He devotes a whole chapter to the subject and is extremely fluent.[18] But his eloquence is pale beside that of Sir Julian Huxley, who has gone a long way beyond his grandfather's view of Darwinism as a provisional hypothesis:

> Two or three states in your country still forbid the teaching of evolution, and throughout your educational system evolution meets a great deal of tacit resistance, even when its teaching is

perfectly legal. Muller, the Nobel Prize-winning geneticist, has written an admirable paper called *One Hundred Years Without Darwin Are Enough,* in which he points out how absurd it is still to shrink from teaching evolution—the most important scientific development since Newton and, some would say, the most important scientific advance ever made. Indeed, I would turn the argument the other way round and hold that it is essential for evolution to become the central core of any educational system, because it is evolution, in the broad sense, that links inorganic nature with life, and the stars with earth, and matter with mind, and animals with man. Human history is a continuation of biological evolution in a different form.[19]

Perfect Faith

There are moments when Simpson euphorically asserts that he and his colleagues have found the ultimate solution to all biological riddles. Examples: "We seem at last to have a unified theory . . . which is capable of facing all the classic problems of the history of life and of providing a causalistic solution of each." "Within the realm of what is clearly knowable, the main problem seems to me and many other investigators to be solved." [20]

Simpson is here referring to the synthetic theory, about which his sober colleague Mayr says: "The basic theory is in many instances hardly more than a postulate." [21] When the euphoria wears off, Simpson himself sees that this is the correct position: "How evolution occurs is much more intricate, still incompletely known, debated in detail, and the subject of most active investigation at present." [22]

Millenarianism

The distinctly religious idea of a heaven on earth is found from time to time in the literature, although it has become a little threadbare in the course of a century. The rasher evolutionists boast that they have the power to control the future. Thus Stebbins, no longer sober and discouraged, concludes his book by declaring (without a shred of evidence): "The control by man of organic evolution is now an attainable goal." [23] It might be prudent to defer such boasts until a blue rose or a black tulip had been produced, but Simpson is bolder still: ". . . it is unquestionably possible for man to guide his own evolution (within limits) along desirable lines." [24] He then goes further and tells the students of St. John's College just what can be done and how we can make ourselves into gods:

> Even now, we know enough about the central process of past evolution, natural selection, to make a good start at improving the breeds of *Homo sapiens,* as we have in fact used this knowledge to improve breeds of other species.
>
> There is, furthermore, reason to think that we are on the verge of further biological discoveries that could make selection far more effective or could even supplant it with other, faster and surer evolutionary processes. It is probable that the incidence of mutations can be controlled within broad limits: instances are known in which the rate of mutation is itself a genetic factor subject to selection. Control over the direction of mutation, possible now only in a few quite special cases, is another eventual probability. Growing knowledge of the actual chemical nature and structure of genes holds the possibility that genes or in the end even whole genetic systems can be made to order. The guidance of evolution could then become a simple matter of following specifications.[25]

I have quoted the last passage to show Simpson's religious fervor, but it would be unfair to let this passage stand alone, because it may not represent his final judgment. In fact, in the years after publishing this passage he must have lost his religion, or there was some alteration in the state of biology, because in 1969 he had quite a different opinion about the future: "The problems of tailoring a gene and inserting it in human sperm or egg, making it hereditary, are so many and so little understood at present that reasonable prediction would place that in a future very remote indeed. Moreover, the human (or any other viable and natural) gene system is so intricately balanced that insertion of a foreign element, however well specified in itself, would probably have disastrous effects." [26]

The reader will by this time have noticed that the atmosphere is full of emotion and inconsistency. At one moment Simpson says that he can improve the breed, and then in a darker mood he states that the control of evolution is a mere dream. This changeableness is not confined to him or even to the American evolutionists. Sir Gavin de Beer confesses remarkable ignorance in one breath: "It must readily be admitted that the causes of the origins of patterns, colors, and of many other things, are not known"; [27] and in almost the next breath asserts that the modern theorists (of whom he is one) have solved everything: "According to the synthetic theory, organic evolution is satisfactorily explained, in principle and in many details, by the mechanisms of heritable variation and natural selection." These are wild swings from arrogance to humility, from boasts of wisdom to confessions of ignorance.

Examples could be multiplied, but I have no desire to extend the list. I would rather make a friendly suggestion, though it is an incongruous one for a layman to offer. It is simply that the scientists should pause, take stock calmly, and work out a consistent position, no matter what it may be. In a nutshell, they should cultivate sobriety.

Lest it be thought that I am cruel to Professor Simpson, let me now say something in his favor. He has refrained from the worst sin. He has not personalized the new religion. He is not anthropomorphic.

When the major aim of Darwin and his followers has been to get rid of all Watchmakers and show that evolution could proceed without divine intervention of any kind, it would be deeply deceitful to smuggle in a Watchmaker in disguise and thereby enjoy the best of two worlds. This has not been done by Simpson or any of the professionals. But Robert Ardrey, the former dramatist and amateur biologist, who regards himself as a faithful exponent of orthodoxy (with a slight African twist), commits this crime, perhaps unwittingly. His words are so poetic that I would like to offer even more of them:

> Never to be forgotten, to be neglected, to be derided, is the inconspicuous figure in the quiet back room. He sits with head bent, silent, waiting, listening to the commotion in the streets. He is the keeper of the kinds.
>
> Who is he? We do not know. Nor shall we ever. He is a presence, and that is all. But his presence is evident in the last reaches of infinite space beyond man's probing eye. His presence is guessable in the last reaches of infinite smallness beyond the magnification of electron or microscope. He is present in all living beings and in all inanimate matter. His presence is asserted in all things that ever were, and in all things that will ever be. And as his command is unanswerable, his identity is unknowable. But his most ancient concern is with order.[28]

Ardrey and Archdeacon Paley are soul-brothers. Paley demonstrates the Watchmaker by processes of reasoning, but these are not essential to Ardrey; he comes to the Man in the Back Room by intuition. By the same token he departs from science. A Jehovah's Witness would say that Ardrey was getting religion, and he would mean old-fashioned religion rather than a scientific brand.

1. Mayr (1963), 8.
2. Simpson (1964), 35.
3. Huxley (1960A), 45–46; but note that Asa Gray (1876), 213, 214, 221, did not believe that Darwin himself intended to deny all design.
4. Simpson (1964), 4; see also viii and 12.
5. Simpson (1949), 132, 145–146, 159, 174–175, 200–202, 246.
6. Deevey (1967), 633.
7. Simpson (1944), 76n; see also (1964), 131.
8. Conklin (1943), 147.
9. Barzun (1958), 63–69, gives an entirely different demonstration of this religious fervor. So does Krutch (1956), 197–198.
10. Hardin (1961), 293.
11. Simpson (1949), 273.
12. Simpson (1949), 272; see also 159 and 273, and Huxley (1942), 473. Simpson seems to recognize that this forces him to speak prematurely, since he says (1944, xviii): "For almost every topic discussed in the following pages the data are insufficient."
13. Simpson (1949), 123.
14. Simpson (1949), 273.
15. Simpson (1949), 277.
16. Simpson (1964), 200.
17. T. H. Huxley (1893), 468–469.
18. Simpson (1964), Chapter 2, 26–41.
19. Huxley (1960A), 42.
20. Simpson (1949), 278–279.
21. Mayr (1963), 8; see also his warning on 7.
22. Simpson (1964), 10; see also (1944), xviii.
23. Stebbins (1950), 561.
24. Simpson (1964), 285.
25. Simpson (1964), 284–285.
26. Simpson (1969), 129; see also 58–59.
27. De Beer (1966), 27, 29.
28. Ardrey (1961), 353–354.

I n examining the single parts of classical Darwinism, I concluded that they were all sadly decayed. How will the biologists receive this statement? Looking at the major parts one by one, it seems probable that my contention as to the breeders' data—that they show micro changes not cumulating into macro effects—will be rejected out of hand by many students, on the ground that they simply *must* cumulate. Many will never go along with me on adaptation for more or less the same reason. As to natural selection, many will concede that they cannot define it, measure it, or even observe it, and that it has tautological elements; but will nevertheless defend it with their hearts' blood. But it seems reasonable to predict that almost no one will defend sexual selection, and that no serious arguments will be made on behalf of survival of the fittest, the struggle for existence, or the two corollaries. Thus large and famous elements of the original doctrine have been discarded, and the total structure presented by Darwin in his lifetime can hardly be endorsed by any modern biologist.* This demonstration meets the demands of my thesis—that the professionals

---

* I must make an exception for Ghiselin (1969), where Darwin is lavishly praised as a thinker, philosopher, and methodologist. Professor Ghiselin would challenge almost every point made in this book, but his position is so intransigent that I cannot regard him as typical of the profession or any substantial part of it.

have moved away from classical Darwinism, even if they cling to neo-Darwinism or the synthetic theory.

Since decayed parts will never make a sound whole, the total theory must also be decayed if we are correct as to the parts. Will the biologists concede this? Experience indicates that most of them, even if they cannot deny, will be reluctant to agree. This would probably always be the case on account of caution or inertia, but at present the air is still full of dust from the jubilees of 1959, when many men were so overcome by their enthusiasm for Darwin himself that they unreservedly renewed their fealty to his theory.* Thus the three volumes of scholarly work collected at the Centennial in Chicago[1] contain hardly a word of doubt or criticism.[2] From time to time we also find students making generous overstatements, such as the following by Professor Bonner of Princeton: "In the hundred years since the publication of *The Origin of Species,* our opinion of Darwin was never so high as it is now." [3]

In the face of such enthusiasm I must admit at once that I can never establish unanimous approval for my revisionist views. But it is not difficult to show that, in moments of frankness, many leading biologists have revealed that Darwin's day is done so far as they are concerned. Often they do this in a meager and piecemeal way, clothing the operation with repeated tributes to Darwin as a man; and such cases will not be fully convincing to the reader. A few writers, however, have not been piecemeal, but have said that they no longer subscribe to classical Darwinism, although the statement has not been loud and unequivocal enough to bring the news to the American public. These men go beyond the passive denial of change; they expressly admit it, sometimes with pride and

* The best illustration of this is the contrast between Mayr (1963), 8, which says humbly that "the basic theory is in many instances hardly more than a postulate," and Mayr (1959B), 10, which makes the extraordinary boast that "no phenomenon has ever been found in organic nature that cannot be interpreted within the framework of the modern, synthetic theory of evolution."

sometimes in despair, sometimes candidly and sometimes only inferentially. I will mention five such men, selected because I have quoted them often in these pages and they cannot be suspected of entertaining any heresy.

Professor Hardin is a very interesting case. He makes the frightening assertion that anyone who does not honor Darwin "inevitably attracts the speculative psychiatric eye to himself," [4] but on the same page he admits my exact point by saying: "True, Darwin is not the last word in science." In private correspondence he assures me that he will never agree that classical Darwinism is obsolete, but at the same time his own works point in that direction. The thrust of the passage I have in mind is veiled (perhaps by intention), but when carefully analyzed I believe that it supports my thesis:

> That the magnificent progress of historical evolution is impossible to the cybernetic process conceived in the most narrowly Darwinian terms has been an intuition of countless minds. Efforts to conceive of other processes have not been wholly happy. Lamarckism, entelechy, *élan vital*, orthogenesis, and the hopeful monster are only a few of the terms associated with the less fortunate efforts. More successful have been Wright's, and Fisher and Ford's proposals that Nature may suspend, as it were, the ordinary laws of accounting—now and then and for a while—during which moratorium improbable new combinations may be thrown together to be tested later. [5]

If I interpret him correctly, Professor Hardin is making three points: 1) Countless minds have perceived that Darwinism has failed. 2) Other explanations have also failed. 3) Fisher, Ford, and Wright may have found the solution in the Sewall Wright Effect (the "moratorium" being the long period of isolation in which genetic drift might occur). Point 1 is enough for my purpose, since any salvaging by Fisher, Ford, and Wright occurred long after Darwin's death, and cannot possibly be counted as part of classical Darwinism. Therefore

I need not remind Professor Hardin that Point 2 is a version of the best-in-field fallacy and is totally irrelevant; and that other biologists feel that it is no longer permissible to mention the Sewall Wright Effect in decent company.[6]

Eiseley is a different case. He has perceived the numerous weaknesses of classical Darwinism, but he stays with it because he hopes to find supporting evidence for it in the future.[7]

As to Professor Deevey, in replying to my article he did not concede that my thesis was correct; indeed, he maintained a pretense of disputing it. But his repeated references to *neo*-Darwinism and *baroque* Darwinism are enough to reveal his thinking;[8] for if classical Darwinism were still in the saddle, there would be no sense in these terms.

The most important testimony comes from Professor Simpson, since he is generally regarded as the dean of modern evolutionists. He takes great pride in the synthetic theory, which is the work of many hands, including those of Fisher, Ford, Wright, Sir Julian Huxley, Waddington, Dobzhansky, Mayr, Stebbins, and Simpson himself.[9] Clearly it is the ruling line of thought today.[10] Equally clearly, it is not the same thing as classical Darwinism. Simpson is explicit: "The full-blown theory is quite different from Darwin's and has drawn its materials from a variety of sources largely non-Darwinian. Even natural selection in this theory has a sense distinctly different, although largely developed from, the Darwinian concept of natural selection." [11] This, I submit, entitles me to say Q. E. D.

My last witness is Sir Julian Huxley. I select him partly because his ancient family connections make his testimony especially telling, and partly because he takes the metamorphosis as a matter of course. He does not argue about classical Darwinism still prevailing; he recognizes at once that it has died and been resurrected in a different form. Nor is there anything grudging about his statement; he is filled with poetic joy as he speaks of "this reborn Darwinism, this mutated phoenix risen from the ashes of the pyre." [12]

Let the reader bear in mind, however, that the large and easy aspect of evolution—the fact that change has taken place and that species have appeared and disappeared—remains untouched even if classical Darwinism is put on the shelf. We say Not Proven to Darwin's suggestion as to how and why, but we do not return to fundamentalism.

This is as suitable a place as any to interpolate a report on a professional controversy that came to my attention only after I had completed the body of this work. The problem is whether evolution will continue in the future or must now be considered exhausted. Presumably most lay readers have never considered the latter possibility, but it has serious support within the biological fraternity.

The leading exponent of the view that evolution is finished was the late Robert Broom (1866–1951) of South Africa. Broom was an outspoken advocate of Watchmakers, but there is no doubt that he was a capable scientist and a candid and highly articulate writer.[13]

Broom asserted that evolution was "practically finished" and that almost every living type of plant or animal was "so specialized that it must either evolve a little further in the same direction, remain unchanged, or die out." [14] In doing this he was fully aware that Sir Arthur Keith and other eminent men disagreed with him and that he would be speedily called to account. Therefore he gave his evidence and his reasons with a good deal of care. His case was strong enough to persuade a number of other students, among whom I was surprised to find Sir Julian Huxley.[15]

The argument is simply that all existing forms are quite specialized; that new types do not arise from specialized forms, which are blind alleys in an evolutionary sense; and that therefore we have no prospects of substantial change in the future. Huxley states it succinctly: ". . . there is no certain case on record of a line showing a high degree of specialization giving rise to a new type. All new types which themselves are capable of adaptive radiation seem to have been produced by relatively unspecialized ancestral lines . . . evolution is thus seen as a series of blind alleys." [16] *

This proposal, if correct, is ruinous to all forms of Darwinian thought, and hence it has been vehemently contested. Simpson is strongly opposed to it.[17] He makes various arguments of a more or less question-begging kind, but the heart of his response is in the following sentence: "Life and its environment are in such ceaseless flux that it is simply inconceivable that a permenent equilibrium will ever be reached." [18] Take note of the phrase *ceaseless flux*. It is the quintessence of Darwinism, as shown in the common attitudes towards species, cumulation of micro changes, and gaps in the fossil record. But is it correct?

Certainly it is not correct for the numerous animals and

* It has slowly dawned on me that substantially the same argument is made by Goldschmidt (1952), 91–92, although Goldschmidt and Broom moved in such widely different circles that they probably never realized their kinship. Goldschmidt says, in sum, that evolution has obviously proceeded from the higher categories to the lower, from the phylum down to the species and subspecies; whereas the Darwinians teach just the opposite —that a subspecies is transformed into a new species by the cumulation of small differences, after which the new species acquires the rank of genus by the long cumulation of further small differences, and so on and on until a new phylum has been created, as the birds are said to have evolved out of the reptiles. Simpson (1953), 350, seems reluctantly to concede this point, but he will not go on to agree with Goldschmidt's conclusion that the process is going in the wrong direction for Darwinian purposes. Simpson sees, perhaps, that this would utterly refute statements such as we find in Mayr (1942), 298: ". . . the origin of the higher categories is a process which is nothing but an extrapolation of speciation." Even if extrapolation were a legitimate practice, it would, according to Goldschmidt, be aimed in the wrong direction. I esteem this because it is a *biological* answer to extrapolation, rather than an exercise in logic or mathematics; but by the same token no outsider can assess it properly. It is a pity that the evolutionists are too deeply committed to give it a fair hearing.

plants that have not changed in any significant way since they first appeared fifty to five hundred million years ago; e.g., the horseshoe crab, the opossum, the oyster, the platypus, the brachiopod called *Lingula,* the ginkgo tree, and the Australian lungfish. No one knows this better than Simpson, who has often tried to explain the persistence of such forms.[19] They are a standing challenge to the hypothesis of ceaseless flux and have defied the explanatory efforts of many famous biologists.[20]

The influence of a powerful dogma on even the strongest mind may be seen in Simpson's occasional attempts to explain the persistent types by contending that their environments have never changed. He says, for example: "Some sorts of environments in the Cambrian and others developing since then have persisted without essential change. Continuity of basic physical conditions may persist for millions of years or may end tomorrow." [21] Thus, when faced with particular anomalies, Simpson is willing to assert that the environment has been stable and unchanging since the Cambrian (the time of the earliest fossils). But, since other plants and animals are part of the environment, and since these others have changed extensively since the Cambrian, this contention is absurd on its face. Fortunately, there is no need to argue this with Simpson, since he knows that it is absurd and sometimes says so very clearly, as for instance: ". . . an environment completely stable for any considerable period of time is inconceivable." [22] The persistent types are a persistent paradox, straining the consciences of the evolutionists.

Only the biologists can judge the ultimate merits of the controversy as to whether evolution is finished.* But every

---

* I can, however, report specious reasoning on the point in England. Hardy (1954) is Sir Alister Hardy's article "Escape from Specialization" and is aimed directly at Huxley, Broom not being mentioned. Hardy seems to agree that the *adult* forms are too rigid to change, but suggests that larval stages could have diverged (pedomorphosis) and that, if this had been combined with a shift of sexual maturity into such stages (neoteny), the trick would be done. Hardy admits this is largely specula-

layman can see how Broom and Simpson are influenced by their views for or against Watchmakers, and how these views drive Simpson, at least, to occasional irrational arguments. On the other hand, one must point out that Nature herself has set the stage for trouble by furnishing evidence of much flux in some fields and no flux in others.[23] If Nature would be more regular and consistent, it would be far easier to construct a sound theory. In the meantime, however, the controversy seems to show the precarious status of the total theory.

I cannot close this chapter without suggesting a splendid opportunity for an ambitious young American biologist. The facet of the total theory that has always most interested the general public is the transition from ape to man, from A to M for short. This is a favorite theme of popularizers, who fly from A to M by waving the wand and thereby give the public the impression that the passage is easy.[24] A bright young man could make his name by telling a different story.

The road from A to M is rough and rocky. It includes the development of language, the achievement of upright posture, and all the other differences of kind or degree between ape and man. I learned how difficult these matters are when I read a recent book in German on the problem of hominiza-tion.[25] It was written by two learned Jesuits, who constructed a strong case by quoting the confessions and recriminations of professional biologists. They had enough material to dispel any feeling that the transition had been explained.

Point M is, of course, familiar to all of us, being the place at which we now stand. But Point A is shrouded in mist. I may even have misstated the problem by assuming that A is

---

tion, and his colleagues seem to have rejected it. Thus Simpson (1949), 327, ignores Hardy's effort, while Mayr (1960), 351, without naming Hardy, repudiates his entire line of thought: ". . . the attempt to 'explain' genetic and selective processes by all sorts of fancy terms like 'pedogenesis,' 'palingenesis,' 'proterogenesis,' and whatnot have had a stultifying effect on the analysis. The less said about this type of literature, the better."

in the kingdom of the apes, since we must always remember that there are no family trees any more and that no one is sure about the ancestral forms. Simpson has a curious passage on this subject: "Apologists emphasize that man cannot be a descendant of any living ape—a statement that is obvious to the verge of imbecility—and go on to state or imply that man is not really descended from an ape or monkey at all, but from an earlier common ancestor. In fact, that common ancestor would certainly be called an ape or monkey in popular speech by anyone who saw it." [26] This reminds one of the learned critic who asserted that the author of the *Odyssey* was not Homer but another poet with the same name. The problem deserves fuller treatment.

Another phase of the A-to-M transition is whether Point A was in the trees or on the ground. All laymen have assumed for a century that man at one time lived in the trees; not precisely as Tarzan did perhaps, but in somewhat the same way. I have recently discovered that the biologists are by no means sure of this.[27] Our young man could do a great service by nailing down this one point.

1. Tax (1960).
2. Olson (1960) is the only exception I have found.
3. Bonner (1962), 45.
4. Hardin (1961), 216.
5. Hardin (1961), 282.
6. Mayr (1963), 204; Goldschmidt (1952), 93; Simpson (1953), 355.
7. Eiseley (1958), 119, and (1961), 320.
8. Deevey (1967), 635, 637, 640.
9. Simpson (1949), 278.
10. Mayr (1963), 7–8, has the modesty to say: "The fact that the Synthetic Theory is now so universally accepted is not in itself proof of its correctness." And, more damningly: "The basic theory is in many instances hardly more than a postulate." A postulate is a position assumed without proof.
11. Simpson (1949), 278n.; see also Mayr (1959B), 4–5.
12. Huxley (1942), 28. See also Huxley (1942), 8, 27; and Simpson (1949), 226; (1953), 58; and (1964), 4. 63–84, and 206.
13. Ardrey (1961), 177–182; Moore (1953), 282–304; Simpson (1949), 277.
14. Broom (1933), 11–14.
15. Broom (1933), 14; Simpson (1949, reprinted 1967), 326; Huxley (1942), 562, 571.
16. Huxley (1942), 562.
17. Simpson (1953), 306–310, 327 ff.; and especially (1949), 326 ff.
18. Simpson (1949), 327; but compare his oracular statement on 249: "Evolution does not proceed from the general to the particular but from the particular to the particular."
19. Simpson (1944), 125–126, 136–141; (1949), 101, 192–195.
20. Ames (1939), 184; Huxley (1957), 111, quoting his grandfather Thomas H. Huxley; Robson (1928), 45–47; Stebbins (1950), 518; Wright (1965).
21. Simpson (1949), 195; see also (1944), 141.
22. Simpson (1953), 184; see also (1944), 188; (1949), 190, 205; (1953), 293; and Mayr (1959B), 9.

23. Mayr (1942), 217.

24. Conspicuous examples are Gamow and Ycas (1968), Morris (1967), and certain publications of the *Life* Nature Library.

25. Overhage and Rahner (1963).

26. Simpson (1964), 12.

27. I began to collect references only recently, but can offer the following mixed bag: Ardrey (1961), 248; Corner (1954), 45; Hardin (1961), 270; Huxley (1942), 570; F. Wood Jones in Eiseley (1957), 101–102; Mayr (1959B), 5, and (1963), 626–630; Smith (1958), 245; and Zuckerman (1954), 301.

The phenomena of biology can be extremely interesting. The work of Konrad Lorenz and his colleagues, for instance, has touched thousands of hearts. Our wonder does not cease as we learn more about the animals and plants; indeed, there seems to be no end to the color and diversity in nature. It is refreshing and profitable to look at the phenomena of life with the unreasoning wonder of a child.

Professor Simpson, however, says it is not enough to look at things with the unreasoning wonder of a child.[1] He and many other evolutionists think they should explain why the phenomena are as they are, and presumably they think that we should listen. But why should they explain? Has anyone really asked any questions? Did anyone ask Robert Ardrey to explain the prairie-dog kiss? Did any mature person ask Professor Tinbergen to explain the patches of colored skin on the female baboon? Nobody asked Simpson to explain the ear tufts on his favorite squirrels, yet he was sorely tempted to try an explanation.*

Sir Julian Huxley inadvertently emphasizes a slightly different aspect of the problem when he speaks of the alternatives

---

* Simpson (1964), viii–ix, is very revealing. He admits that "it is irresistible to speculate and to extrapolate." He seems to think that this sort of activity should "be taken in a spirit of good clean fun."

to confessing ignorance: ". . . if we repudiate creationism, divine or vitalistic guidance, and the extremer forms of orthogenesis, as originators of adaptation, we must (unless we confess total ignorance and abandon for the time any attempts at explanation) invoke natural selection." [2] Socrates might have asked why we should not confess ignorance. Is there a moral duty to offer an explanation, especially when we know that it is weak? Why use the word *must*?

These questions are important because the passion to explain has injurious consequences. First, as all my authors know, some explanations are so feeble that they sound more like the first thoughts of a freshman than the insights of a sage.[3] They make the profession look foolish. Professor Bonner recognizes this when he says: "The answers may come with further study, but they must be discovered by physiological experiments, not by complacent speculation." [4] We must reach a point where warnings not to indulge in complacent speculation will no longer be necessary.

Second, all this explaining takes the charm out of biology. The subject becomes dull. In place of the delights of nature we are offered the deserts of argumentation. This leads to painful experiences when evolutionists address gatherings of high-school teachers; they learn that the teachers have no questions and want no answers. Simpson reports: "As regards my subject, evolution, a significant minority of them simply do not believe a word of it and automatically close their minds when the subject is named." [5] Simpson considers this unreasonable, but how many people find lectures on evolution lively or profitable?

I have no objection to scientific inquiry into the vast stretches of the past, but I have come with time to regard it as an entertaining pursuit for the researcher rather than as a moral obligation for either the scientists or the public.[6]

I also have no objection to explanations, if they are good

explanations. Unfortunately, in the field of evolution most explanations are not good. As a matter of fact, they hardly qualify as explanations at all; they are suggestions, hunches, pipe dreams, hardly worthy of being called hypotheses. A man who was careful and precise in his language would never have used the word *theory* for classical Darwinism when it could not explain extinction, the eye, hypertely, and many other major problems. It was an interesting suggestion for Darwin's colleagues to debate, but it should have been seasoned for a long time before being ranked as a theory.*

This is not mere quibbling. The profession has worked itself into an embarrassing position when Sir Julian Huxley tells the television audience: "The first point to make about Darwin's theory is that it is no longer a theory, but a fact," [7] while at almost the same time Professor Mayr, addressing himself to serious students, says: "The basic theory is in many instances hardly more than a postulate." [8] Such an enormous discrepancy between two leaders (both of whom were really talking about the synthetic theory rather than Darwin's own) is bad for the standing of the profession. The public may rightly feel that it has been paltered with if the complexity and insecurity of the theory are not laid before it.

As my final constructive contribution, I would like to suggest that some of the effort devoted to explaining should be diverted to contemplating, and I will even offer a theme that might be fruitful for contemplation. We have seen that family trees, though always expected by laymen, are no longer furnished by cautious biologists. In the days when they were still in use, man was always at the apex of the tree; and he will doubtless be there again if trees are ever revived, since neither the biologists nor the laymen have ever seriously thought of

* The elder Huxley expressly called it a hypothesis and recommended that it be held "with a light hand" and only provisionally: T. H. Huxley (1893), 468–469.

any other possibility. But when we put man at the apex, we are overlooking the admitted fact that, anatomically speaking, man is primitive rather than advanced. As Robert Ardrey says: "From the point of view of evolution, therefore, it is the specialized animal that must be regarded as the more advanced; the animal retaining his generality, the more primitive. Unless we grasp this concept—that man, for instance, is on the whole a more primitive creature anatomically than the gorilla—then we shall have difficulty in tracing the human emergence through the obscure landscapes of our antique past." [9]

This consideration, startling as it is, should be connected with an obvious but even more startling aspect of the human hand. Let us for a change, if only as a stimulating exercise, turn our sights around and try to *devolve* or despecialize the basic forelimb of the vertebrates. What do we find? The foot of an antelope, the wing of a bat, the flipper of a sea lion, and all the other multitudinous forms known to comparative anatomy "can be reduced to a common plan—one bone in the upper arm, two in the lower, a number in the wrist, and five fingers. The variations are brought about by the enlargement of some parts, as with the bat's fingers, the reduction or loss of others, as in the side toes of the antelope, or the joining of originally separate parts into one, as in the antelope's cannon bone. The plan is the same, though one is used for running, one for grasping, one for flying, and one for swimming. This only has any meaning if the different creatures are all descended from a common original ancestor possessing this plan for fore-limb structure in a simple and primitive form, and that then, in the course of evolution, they specialized in different directions." [10]

Sir Julian Huxley lets the matter rest at this point, but if we are looking for a form possessing the forelimb structure in a simple and primitive form, perhaps it is staring at us in the human hand. I will not assert that all vertebrates are descended from man, and I will not guarantee rich results from this line of thought, but if a few biologists would brood on these strange facts, testing them and expanding them in various ways, they might find new insights occurring to them.[11]

If I were attacking Darwin's character, as Samuel Butler did, it would only be natural for his admirers to rush to his defense. If I were impugning his intelligence, as Barzun sometimes does, again it would be natural for them to protest. But I am asserting only that he is not the last word in science and that his theory has not been proved adequate. This is not a proposition that should stir the emotions rather than the reason.

Darwin was an amateur.[12] He did not teach in a university or work in a laboratory. He "did" science in his own house with no trained staff and very little equipment. He worked a maximum of four hours per day. In his time there were very few full-time biologists in all of England.

Nowadays we have several thousand professional biologists in the United States alone. They have thorough training, great skill in many techniques, large laboratories, fine equipment, clerical help. They work, I assume, a good deal more than four hours per day.

It would be strange if this army of well-equipped professionals had not been able, in the course of eleven decades, to go beyond the hypothesis worked out by a lone amateur in 1859. If Hardin and Deevey are correct in saying that classical Darwinism is still the ruling paradigm, we must ask what all these men have been doing. If I am wrong in saying they have worked out something new, they should blush with shame. When they contend that classical Darwinism is still in the saddle, they are stultifying their own profession. Therefore I ask—Must we defend?

Any profession that does not supply its own criticism and

iconoclasm will discover that someone else will do the job, and usually in a way it does not like. This is already occurring with the evolutionists, as witness Anthony Standen's paperback *Science Is a Sacred Cow* (Dutton, 1950). Its ten raucous pages on Darwinism must have destroyed a good deal of confidence in the intelligence and integrity of the exponents of the theory.

It is my conviction, after examining the literature, that intelligence and integrity are still very much alive among the biologists. In their own circles they speak candidly and express their misgivings freely. Only when they popularize do they become pompous and pontifical.[13] Perhaps they are reluctant to confess error. Perhaps they fear that the fundamentalists will gloat over their discomfiture. These would be human failings, but just the sort that one must resolutely put aside. I urge the Darwinists to take the public into their confidence by a full disclosure. They are not expected to be infallible, confession is good for the soul, and candor is always highly valued.

1. Simpson (1949), 273; and (1969), 11. Medawar (1960), 102–103, takes a much more moderate position.
2. Huxley (1942), 473.
3. Mayr (1961), 1503; McAtee (1932), 1–3; Simpson (1964), 202–204; Smith (1958), 15–25; Stebbins (1950), 101–104.
4. Bonner (1962), 151; but on 175 Bonner forgets his own admonition.
5. Simpson (1964), 30.
6. Barzun (1964); cf. Simpson (1964), ix, on "good clean fun."
7. Huxley (1960A), 41; see Olson (1960), 506, for a critique of this kind of language.
8. Mayr (1963), 8.
9. Ardrey (1961), 252–253; see also Simpson (1949), 251.
10. Sir Julian Huxley as quoted in Clark (1968), 259–260; see also Ardrey (1966), 251.
11. Simpson (1969), 115–116, touches on a parallel phenomenon in the realm of language—that the simpler languages seem to be *less* primitive, so that language becomes more complicated as one goes back in time. He cites English and Chinese as actually doing this, but instead of pursuing the matter further he abandons it quickly, saying: "Yet it is incredible that the first language could have been the most complex." My suggestion is that it pays to examine what seems incredible; see Olson (1960), 542–543.
12. Hardin (1961), 33, 38–39, is my source for the particulars on Darwin.
13. Williams (1966), 55, sees things in very much the same way: "It is mainly when biologists become self-consciously philosophical, as they often do when they address nontechnical audiences, that they begin to stress such concepts as evolutionary progress. This situation is unfortunate, because it implies that biology is not being accurately represented to the public."

The Goldschmidt Case

Richard B. Goldschmidt (1878–1958) of the University of California was a first-class geneticist. Even in the heat of battle his major opponent, George G. Simpson of Harvard, paused for a moment to state that among Goldschmidt's peers there was "profound respect and admiration for his genetical work." [1] He could not be derided as an amateur, outsider, vitalist, or Lamarckian. His credentials were impeccable.

It was in 1940, shortly after moving from Berlin to Berkeley, that Goldschmidt published his major work, *The Material Basis of Evolution*. This touched off a controversy that has not yet entirely died down. The public hardly knew that anything was going on, but within the profession the Goldschmidt episode was a much greater event than the Scopes trial. The thrust and parry are worth reconstructing because they show how far we have moved from Darwin and how shaky the structure of evolutionary theory really is.

I have been unable to find any summary, no matter how partisan, of the arguments. Comments on various points are plentiful, but there is no systematic study of the whole. There was never a public debate in which the issues could be threshed out one by one. Goldschmidt put out his book. There were reviews. There were discussions and partial refutations

by other authors. There were a few supporting remarks by
Goldschmidt's friends. There was one more article by Goldschmidt himself. But there was never a summing-up or a thorough evaluation.

This has led me to make my own analysis, which is in the
form of a conversation between four leading figures in the
profession, G. G. Simpson, Ernst Mayr, Sewall Wright, and
Goldschmidt himself. No such conversation ever actually occurred, and it is full of anachronisms because it sets forth
opinions that were never expressed on a single occasion or
in a single place. It distills the views of each man and presents
them in their proper relation to the whole, and it tries to do
this fairly.

Imaginary Conversation

Goldschmidt: Gentlemen, I have been reviewing the present condition of evolutionary theory and I have concluded
that we are in serious trouble. It is perfectly clear that many
species and genera, indeed the majority of them, appear suddenly in the record, differing sharply and in many ways
from earlier groups. This apparent discontinuity becomes
more common the higher the level, until it is virtually universal as regards phyla, classes, orders, and even families, all
the higher stages in the taxonomic hierarchy.[2] Some students
are not disturbed by this, but to me and to a number of
friends who are inclined to take the record at face value, it
is extremely disturbing.

At the same time the geneticists are talking more and more
about little "point mutations" as the effective agents in caus-
ing change. Simpson here even says that their importance
varies inversely with their size,[3] and one of my colleagues in
California says they are really in the domain of the nuclear
people.[4] Thus we have big problems and small answers,
which is what leads me to say that we are in trouble.

Wright: These are old problems, well known to all of us.
Why do you suddenly feel that we are in trouble?

Goldschmidt: Because after forty years of working with

micromutations, I am ready to give up on them. They seem
to lead nowhere.

Mayr: What do you call micro?

Goldschmidt: I cannot give you a definition or a clear line
of demarcation. It is merely large and small. Everything in
the genetics lab is micro.[5] Everything in the breeding pens is
micro. David Lack, with his finches, is micro.[6] I cannot give
concrete examples of macro because I have never seen any.

Simpson: Why should you give up? The little changes
cumulate into big ones.

Goldschmidt: I cannot believe that any more. If you combined a thousand mutations into one fruit fly, which is
mathematically impossible, it would still be a fruit fly.[7]

Mayr: I am an ornithologist, so David Lack means a lot
to me. How can you brush him off so lightly?

Goldschmidt: All he found on his Galápagos expedition
was differences in beak and feather, although the birds are
supposed to have been isolated for thousands of years. If you
will look at *The Origin of Species,* which you have just
persuaded Harvard to reprint in facsimile for the convenience of the profession, you will find Darwin saying: "That
most skilful breeder, Sir John Sebright, used to say, with respect to pigeons, that he would produce any given feather
in three years, but it would take him six years to obtain head
and beak." [8] So this is all old stuff. We are making trips to
the Galápagos to prove things that Darwin found out at
home. We are getting nowhere.

Wright: What do you propose to do about the situation?

Goldschmidt: That is what I wanted to explain today,
partly because all of you should know about it and partly
because I would like to get your reactions. We need something larger than point mutations, so I want to suggest that
there must have been what I will call "systemic mutations,"
something that shook up the whole system but still allowed
it to survive and breed. It must have been a change of intra-
chromosomal pattern. Genes and gene mutations would not
enter into it at all.[9]

Simpson: I suppose you would agree that there is no direct
evidence for this. After all, no one has ever observed such a
mutation.

Goldschmidt: Of course no one has. I only say that it *must* have occurred.

Simpson: And it would require a wide departure from ordinary gene theories? [10]

Goldschmidt: Certainly, although there are lines of evidence pointing in this direction. Position effects, for example.

Wright: It seems to me that a big mutation such as you suggest would be so difficult to assimilate that it would probably be fatal. [11]

Goldschmidt: You are right. I only say it might succeed one time in a thousand.

Mayr: A big shakeup like that might also produce a monster. Isn't that about the best you would get?

Goldschmidt: Yes, but it would be a hopeful monster, in the sense that its divergences gave hope for a glorious future.

Mayr: The suggestion is attractive, or seductive, in that it would solve a lot of problems. But in the present state of the evidence I do not think you could call it a theory, or even a hypothesis. It is more in the pipe dream stage.

Simpson: I am of the same opinion. But I want also to challenge your premises. You speak of the gaps as though they were enormous and unbridgeable, but the situation is really not that bad. The record is not utterly devoid of transitions. They are common between species and genera, and there are some between classes, although they get pretty rare above that. But we have links now between fishes and amphibians, amphibians and reptiles, reptiles and birds, reptiles and mammals. [12]

Goldschmidt: You always take that rosy view, but you know how meager your evidence is. You have hardly an indication of intermediate forms in those pairs you mention. [13]

Wright: Gentlemen, please do not start this old battle all over again. We all know that the gaps look easy to optimists and hopeless to pessimists. I have wished for forty years that the paleontologists would reach some kind of consensus on them, since we mathematicians cannot work with any confidence when you cannot give us solid premises. [14]

Goldschmidt: Very well, I will switch from phyletic gaps,

which are really discontinuities in family trees, to the emergence of evolutionary novelties, by which I mean the first appearance of new things, such as blood or lungs. You do not have to trace ancestors; you only have to show me how these things could grow up on a step-by-step basis.

Simpson: I have just done a neat study on the first appearance of crochets, the little spurs or crests on the cheek teeth of horses.[15]

Goldschmidt: This is hopelessly micro. I mean big things. I have made a list of seventeen, and I challenge you to explain any of them on a step-by-step basis.[16] Here they are: hair in mammals; feathers in birds; segmentation of arthropods and vertebrates; the transformation of the gill arches in phylogeny; teeth; shells of mollusks; ectoskeletons; compound eyes; blood circulation; alternation of generations; statocysts; ambulacral system of echinoderms; pedicellaria of the same; cnidocysts; poison apparatus of snakes; whalebone; and primary chemical differences like hemoglobin versus hemocyanin. I could give you many more from the plant world, but since you are not botanists I will spare you.

Mayr: I have done a little work on the birds known as Hawaiian honeycreepers.[17] Would that be useful here?

Goldschmidt: No, that is just beaks again. You would be on the right track if you worked out an explanation of the wing shape, type of flight, and correlated structure of the lungs of a hummingbird, together with honeysucking bill and tongue.[18]

Mayr: I have done a lot of work on novelties and I think we can find step-by-step solutions.

Goldschmidt: I have read your article. It is a good try, but it is too general. Look at the famous old problem of the eye. You say: "The evolution of the eye ultimately hinges on one particular property of certain types of protoplasm—photosensitivity. This is the key to the whole selection process. Once one admits that the possession of such photosensitivity may have selective value, all else follows by necessity."[19] This is really dealing in broad strokes. You ought to discuss retina, cornea, rods and cones, visual purple, and all kinds of details. At the same time I can understand your leaving them alone, because I think it is impossible to explain them.

Simpson: But if we went along with you, the innumerable studies of microevolution would be unimportant. They would hardly have any value in the study of evolution as a whole.[20]

Goldschmidt: They would have a little value in explaining how the details were worked out after the macro changes took place,[21] but in general you are correct. I have been a geneticist all my life, but I must confess that a lot of time has been wasted in our laboratories.

Simpson: I am not going to go along with you. I am convinced that little-by-little will do it.

Goldschmidt: You are wrong. It is not only that the little mutations do not cumulate; they are also going in the wrong direction. You yourself say: "Evolution does not proceed from the general to the particular but from the particular to the particular." [22] That simply will not explain the higher categories. We need something that creates the general, not a new subspecies or variety.

Mayr: But that is silly. The higher categories must have started as species, even if they rose to genera, et cetera, later as their progeny multiplied and diversified.[23]

Goldschmidt: Oh no. Surely the higher categories were first in time as well as in the classifications.

Simpson: I think Mayr is right. They were certainly species when they first appeared, no matter what happened later.[24]

Goldschmidt: You amaze me. Mine is the natural, naïve view, but I do not see how it can be wrong.[25] If it is, what happens to all your phylogenies and your calculations of tempo?

Wright: Take it easy, my friends from Harvard. There is a lot of truth in what Goldschmidt is saying. Evolution certainly works down from the higher categories to the lower, rather than the reverse.[26]

Goldschmidt: Thank you, Sewall. It is pleasant to receive a vote of confidence on any subject today.

Simpson: You may be correct, but only in the chronological sense.

Goldschmidt: That is the only sense involved in my statement.

Simpson: I have to repeat that this idea of the hopeful monster is pure speculation. It is terrible to think of what it does to all genetic theory.

Goldschmidt: The time has come to reshape all genetic theory anyhow. The classical theory of the gene as an actually existing unit, lying on the chromosome like a bead in a string of beads, is no longer tenable.[27] We cannot focus on genes and loci, or even on chromosomes, as we always have. Something bigger controls the whole system. That is why I speak of systemic mutations.

Mayr: We are certainly due for some drastic revisions in our concepts of genes, but I doubt if they will have much effect on evolutionary theory.[28]

Simpson: I admit that genes and mutations are still pretty mysterious, but you are asking too much and offering too little.[29]

Goldschmidt: It is a bitter pill, but we are in a desperate position. It was not easy for me either.

Simpson: Again I say that I cannot go along with you. There is no sense in trying to analyze your macroevolution in detail, since you have not been able to put much flesh on the bare bones. I will merely take the position that we need not discuss your hypothesis if there is another that is equally or more probable.[30]

Goldschmidt: That sounds reasonable, but is there such another hypothesis?

Simpson: There is what we were brought up on—minor changes cumulating to work out macro problems step by step.

Goldschmidt: That brings us right back to the gaps. How are you going to handle them?

Simpson: I admit that the gaps are hard problems.[31] They always have been. But we can approach them as our fathers did.

Goldschmidt: How is that?

Simpson: First, we can take refuge, as Darwin did, in "the extreme imperfection of the geological record" and "the poorness of our paleontological collections."[32] The fossils are very spotty, but occasionally we find a new one that seems to fit into a time gap, and it is always an intermediate form in structure.[33] This gives me hope that we will find more, and ultimately be able to close the gaps.

Goldschmidt: Greater faith hath no man. I admit that there have been some finds, but they were very meager. It is like a man who can jump ten feet and is faced with a hundred-foot gap. You do not help him much with one find. He needs nine, and they must be very well spaced.

Simpson: But there is ample reason to believe that the transitional forms simply were not registered as fossils. After all, they were small populations and were probably evolving pretty fast.

Goldschmidt: Are you coming to what is known, with apologies to our friend here, as the Sewall Wright Effect?

Simpson: Yes, the idea that the rules of selection are relaxed in a small population, so that unusual things can follow if a few mutations occur. Of course, the normal result is extinction, but it could be a big change if all went well.[34]

Wright: My name is now irrevocably attached to this thing, but I never intended it to be a panacea. I merely tossed it out as a suggestion.

Goldschmidt: It is used now like a *deus ex machina* whenever a hard problem comes up.[35] Look at my colleague Garrett Hardin for example. He speaks of Wright's proposal "that Nature may suspend, as it were, the ordinary laws of accounting—now and then and for a while—during which moratorium improbable new combinations may be thrown together to be tested later. Wright works the miracle by the errors of small numbers." [36] Sewall, my old friend, do you recognize your brain child?

Wright: I am always embarrassed by such exuberance. As a matter of fact, I have thought of abandoning the whole idea.

Mayr: I have noticed that applying a technical term such as "the Sewall Wright Effect" has a peculiarly soothing effect on the human mind. Among the many inappropriate uses of this effect as an "explanation" of evolutionary phenomena, none is so farfetched as its use to interpret gaps in the fossil record.[37] I do not want to be hard on you, George, but you really have been a little too free and easy with this device.

Simpson: If that is the way you all feel, then I must stand corrected;* but it was all done in a good cause.

Goldschmidt: This seems to leave us with the gaps unsolved. Do you agree?

Simpson: I will never agree. My whole life depends on finding a solution.

Goldschmidt: Maybe there is a solution in my systemic mutations. We can call it macroevolution and use it only for the higher categories.

Simpson: I am going to make a terrible admission.[38] There are really three separate modes of evolution, although we have spoken of it as a single process heretofore. One is *speciation,* the forming of new varieties and subspecies by means of the "point mutations" that the geneticists work with.[39] Then there is a bigger group of changes to be seen in the fossils, especially in the stock example of the horses. This is the sort of thing the paleontologists find, and they arrange them in phylogenies or family trees. We can call this process *phyletic evolution.*[40] I have always regarded these two processes, speciation and phyletic evolution, as sufficient to explain the facts, although I have been uneasy at times. I cannot go along with Goldschmidt, but nonetheless there is a difference, and many of the major changes cannot be considered as simply caused by longer continuation of the more usual sorts of minor changes.[41] I have to admit also that they occur with unusual rapidity, although I will never agree that it is all done instantaneously in one step. I do not know whether these cases differ in kind or only in degree from speciation and phyletic evolution, but I am going to assume that it is a matter of degree until shown otherwise.[42] I will not accept Goldschmidt's invitation to join him in calling this macroevolution. In my next book I will call it *quantum evolution* and will list it as the largest of the three modes of evolution.[43]

---

* Simpson (1964) no longer speaks of the Sewall Wright Effect or of genetic drift, but presents the same discredited idea under the name of "sampling errors" on 20, 73, 74, 75, and 211; although he confesses (75) that "the present consensus is that it is usually overbalanced by selection, that is, it rarely leads to elimination of a genetic factor favored by selection or fixation of one opposed by selection."

Goldschmidt: My friend Petrunkevitch, the spider king at Yale, would say you were adopting my idea.[44]

Simpson: I cannot clearly distinguish between what you call macro and what I call quantum. We are both just groping, and the outlines are not firm enough for a comparison. You may think I am stealing your idea, and that may turn out to be true; but I doubt if it has originality in any event, and I simply cannot adopt it as it is when you have identified it with wild dreams like systemic mutations and the hopeful monster.

Goldschmidt: How are you going to make it work without these things?

Simpson: I do not know. I may not even try to show how it works.[45] This would not be unreasonable when I don't know myself.

Goldschmidt: What will you use as an example?

Simpson: I was thinking of hypsodonty in horses, a very important change in their dentition.[46]

Goldschmidt: That seems pretty micro. Why not take one of my seventeen big ones?

Simpson: I am going to leave those to Mayr.

Goldschmidt: George, you are a remarkably well-informed man and you have been very patient with me today, but we should all recognize that you are constitutionally unable to deal with discontinuities and apparent saltations. You are allergic to gaps and leaps. For instance, Robb shows pretty clearly that there was a big one-step change in the horse's foot when the lateral digits suddenly went down to their present size. You know this field well and you admit that there are no intermediate stages among the fossils, yet you cannot accept what the record shows. You say there must be missing fossils, you suggest that Robb was not very thorough, and you assert that even if it is all true it does not amount to much in the way of a saltation.[47] Sometimes you make me wonder whether you have an open mind.

Wright: Don't badger him, Richard. We were all brought up to fight discontinuities and saltations. Darwin did it and we have followed. If we admitted them, all our theories would go to pieces. I don't know who first said that nature did not make leaps (*Natura non facit saltum*), but it was already an old maxim in Darwin's day.[48] We were all weaned on it, and you will not get us away from it if we stay here all night. George may be a little blunter than the rest of us, but we are all equally pigheaded.

Mayr: That is true. I will fight to my dying day against your idea of novelties appearing suddenly. It *must* have been step by step. I will not go along with Simpson if he speaks of quantum evolution, however vaguely, and many of our colleagues will feel the same way about it.[49]

Goldschmidt: Ernst, you are putting yourself in a weak position. I have just been looking at your big book, where you say: "The development of the evolutionary theory is a graphic illustration of the importance of the *Zeitgeist*. A particular constellation of available facts and prevailing concepts dominates the thinking of a given period to such an extent that it is very difficult for a heterodox viewpoint to get a fair hearing. Recalling this history should make us cautious about the validity of our current beliefs. The fact that the synthetic theory is now so universally accepted is not in itself proof of its correctness." [50] Don't you think that after writing that you ought to be a little cautious and listen very carefully to opposing views?

Mayr: I am, as Milton said, unmoved, unshaken, unseduced, unterrified. Systemic mutations and the hopeful monster are forever *verboten*.

Wright: So be it. Let us part in peace.

This conspectus does not give Goldschmidt a victory, which would be impossible when his idea was obviously no more than a hunch. It does convey a better impression of his case than is common in the literature, but this is no more than fair when the negative aspects of his case were well founded. In the usual treatment there is a more or less careful refutation of the hopeful monster, but no discussion of why Goldschmidt felt it necessary to propose such an outrageous idea. Thus Hardin says: "This important geneticist believed, for reasons that are not at all clear to his colleagues, that . . ." etc.[51] Yet

Goldschmidt was clearly suggesting rather than believing, and he made his reasons perfectly clear to anyone who was willing to listen. I have not had to improve on his presentation, which was always lucid and vigorous. My function was only to gather the admissions scattered through the texts, and to direct attention to the background rather than concentrating on the hopeful monster. The evidence had to be assembled, but not embellished.

1. Simpson (1953), 85n.
2. Simpson (1944), 99.
3. Simpson (1953), 93.
4. Stebbins (1950), 85.
5. Huxley (1942), 153; Simpson (1944), 202.
6. Lack (1947).
7. Goldschmidt (1952), 94.
8. Harvard University Press (1966), 31.
9. Goldschmidt (1940), 206.
10. Simpson (1949), 230–239.
11. Fisher (1930), 128.
12. Simpson (1949), 230–239.
13. Goldschmidt (1952), 92.
14. Goldschmidt (1940), 137, 169; Huxley (1942), 151–152.
15. Simpson (1944), 59–60; (1953), 105–106.
16. Goldschmidt (1940), 6–7.
17. Mayr (1963), 590–591.
18. Goldschmidt (1952), 93.
19. Mayr (1960), 359.
20. Simpson (1944), 97.
21. Goldschmidt (1940), 206.
22. Simpson (1949), 249.
23. Mayr (1963), 600–601.
24. Simpson (1953), 237–238, 342–346.
25. Goldschmidt (1952), 91–92.
26. Simpson (1953), 350.
27. Goldschmidt (1940), 247.
28. Mayr (1963), 170–172.
29. Simpson (1949), 216.
30. Simpson (1944), 115–116.
31. Simpson (1944), 106–108.
32. Darwin (1859), 280, 287.
33. Simpson (1944), 105–124.
34. Simpson invokes the Sewall Wright Effect frequently. Thus the index to his 1953 book shows it four times under the heading "genetic drift"; see especially page 120.

35. Goldschmidt (1952), 93.
36. Hardin (1961), 282.
37. Mayr (1963), 204.
38. Simpson (1949), 230–239.
39. Simpson (1944), 202.
40. Simpson (1944), 206.
41. Huxley (1942), 153; Simpson (1949), 235.
42. Simpson (1944), 97.
43. Simpson (1944), 197–217.
44. Petrunkevitch (1952), 102.
45. This is the way it is handled in Simpson (1944, 1949, 1953).
46. Simpson (1944), 206–210.
47. Simpson (1944), 61.
48. Darwin (1859), 194.
49. Quantum evolution is mentioned in a footnote on page 310 of Eiseley (1961), but not at all in Hardin, Huxley (1957), Mayr (1963), Rensch, Smith, or Stebbins, although all these books contain abundant references to Simpson's works.
50. Mayr (1963), 7.
51. Hardin (1961), 226.

Abercrombie, M., C. J. Hickman, and M. L. Johnson, 1966, *A Dictionary of Biology* (Penguin, 5th ed.).

Agassiz, Louis, 1860, "Professor Agassiz on the origin of species," *American Journal of Science and Arts*, 2nd series, XXX (July), 142–154.

Allee, W. C., 1955, "Biology," in J. Newman, ed., *What Is Science?* (Simon & Schuster), 243.

Ames, Oakes, 1939, *Economic Annuals and Human Cultures* (Botanical Museum of Harvard University).

Ardrey, Robert, 1961, *African Genesis* (Dell); 1966, *Territorial Imperative* (Atheneum Publishers).

Barker, A. D., 1969, "An approach to the theory of natural selection," *Philosophy* XLIV, No. 170, 271–289.

Barzun, Jacques, 1958, *Darwin, Marx, Wagner* (Doubleday, revised 2nd ed.); 1964, *Science, The Glorious Entertainment* (Harper & Row).

Bateson, William, 1894, *Materials for the Study of Variation* (Macmillan, London); 1922, "Evolutionary faith and modern doubts," *Science* 55: 55–61.

Bonner, John Tyler, 1962, *The Ideas of Biology* (Harper).

Brooks, John Langdon, 1957, "The species problem in freshwater animals," in E. Mayr, ed., *The Species Problem* (American Association for the Advancement of Science, Publ. No. 50), 81–123.

Broom, Robert, 1933, "Evolution—is there intelligence behind it?" *South African Journal of Science* 30: 1–19.

Butler, Samuel, 1879, *Evolution, Old and New* (Hardwicke and Bogue, London).

Clark, R. W., 1968, *The Huxleys* (McGraw-Hill).

Conklin, Edwin G., 1943, *Man Real and Ideal* (Scribner).

Corner, E. J. H., 1954, "The evolution of a tropical forest," in Huxley, Hardy, and Ford, 34–46.

Darwin, Charles, 1859, *The Origin of Species* (John Murray, London; facsimile printed by Harvard University Press, 1966); 1871, *The Descent of Man and Selection in Relation to Sex* (Modern Library).

De Beer, Sir Gavin R., 1966, "In the genes," *New York Review of Books* 6: 14 April, 27–32; 1970, "The evolution of Charles Darwin," *New York Review of Books*: 17 December, 31–75.

Deevey, Edward S., Jr., 1967, "The reply: letter from Birnam Wood," *Yale Review* 61: 631–640.

De Grazia, Alfred, R. E. Juergens, and L. C. Stecchini, 1966, *The Velikovsky Affair* (University Books).

Dobzhansky, Theodosius, 1941, *Genetics and the Origin of Species* (Columbia University Press, 2nd ed.); 1956, "What is an adaptive trait?" *American Naturalist* 90: 337–347.

Dunbar, Carl O., 1960, *Historical Geology* (John Wiley & Sons, 2nd ed.).

Dupree, A. Hunter, 1959, *Asa Gray* (Harvard University Press).

Eiseley, Loren, 1943, "Archeological observations on the problem of post-glacial extinction," *American Antiquity* 8: 209–217; 1946, "The fire-drive and the extinction of the terminal Pleistocene fauna," *American Anthropologist* N.S. 48: 54–59; 1958, *The Immense Journey* (Vintage Books); 1960, *The Firmament of Time* (Atheneum Publishers); 1961, *Darwin's Century* (Doubleday).

Farrand, William R., 1961, "Frozen mammoths and modern geology," *Science* 133: 729–735.

Fischer, David H., 1970, *Historians' Fallacies* (Harper & Row).

Fisher, Sir Ronald A., 1930, *The Genetical Theory of Natural Selection* (citations are to paperback 2nd rev. ed. in 1958, Dover Publications); 1954, "Retrospect of the criticisms of the theory of natural selection," in Huxley, Hardy, and Ford, 84–98.

Gamow, George, and Martynas Ycas, 1968, *Mr. Tompkins Inside Himself* (Allen & Unwin, London).

Ghiselin, Michael T., 1969, *The Triumph of the Darwinian Method* (University of California Press).

Goldschmidt, Richard B., 1940, *The Material Basis of Evolution* (Yale University Press); 1952, "Evolution, as viewed by one geneticist," *American Scientist* 40: 84–98.

Grant, Verne, 1957, "The plant species in theory and practice," in Mayr, 39–80; 1963, *The Origins of Adaptations* (Columbia University Press).

Gray, Asa, 1876, *Darwiniana* (Appleton; paperback reprint by Harvard University Press, 1963).

Greene, John C., 1963, *Darwin and the Modern World View* (Mentor).

Haldane, J. B. S., 1935, "Darwinism under revision," *Rationalist Annual*, 19–29.

Hall, Wilbur, 1939, *Partner of Nature* (Appleton-Century).

Hardin, Garrett, 1961, *Nature and Man's Fate* (Mentor).

Hardy, A. C., 1954, "Escape from specialization," in Huxley, Hardy, and Ford, 122–142.

Hofstadter, Richard, 1955, *Social Darwinism in American Thought* (Beacon Press).

Huxley, Sir Julian S., 1938, "Darwin's theory of sexual selection," *American Naturalist* 72: 416–433; 1942, *Evolution, the Modern Synthesis* (Allen & Unwin, London); 1957, *Evolution in Action* (Mentor); 1960A, "At Random," a television preview on 21 November 1959, in Tax (1960), Vol. 3, 41–65; 1960B, "The emergence of Darwinism," in Tax (1960), Vol. 1, 1–22.

Huxley, Sir Julian S., A. C. Hardy, and E. B. Ford, eds., 1954, *Evolution As a Process* (Allen & Unwin, London).

Huxley, Thomas H., 1893, *Darwiniana* (printed in U.S. by Appleton, 1901).

Kellogg, Vernon, 1925, *Evolution, the Way of Man* (Appleton).

Krutch, Joseph Wood, 1956, *The Great Chain of Life* (Houghton Mifflin).

Kuhn, Thomas S., 1962, *The Structure of Scientific Revolutions* (University of Chicago Press).

Lack, David, 1947, *Darwin's Finches* (Cambridge University Press; paperback ed. by Harper Torchbook, 1961).

Lee, K. K., 1969, "Popper's falsifiability and Darwin's natural selection," *Philosophy* XLIV, No. 170, 291–302.

Lerner, I. Michael, 1958, *The Genetic Basis of Selection* (John Wiley & Sons).

Lyell, Sir Charles, 1853, *Principles of Geology* (9th ed.).

Manser, A. R., 1965, "The concept of evolution," *Philosophy* XL, 18–34.

Mathiessen, Peter, 1967, "The wind birds," *The New Yorker*, 3 June, 42–105.

Matthews, William H. III, 1962, *Fossils, an Introduction to Prehistoric Life* (Barnes & Noble).

Mayr, Ernst, 1942, *Systematics and the Origin of Species* (Columbia University Press; paperback by Dover Publications, 1964); 1957, ed., *The Species Problem* (American Association for the Advancement of Science, Publ. No. 50); 1959A, "Darwin, Agassiz, and evolution," *Harvard Library Bulletin* 13: 165–194; 1959B, "Darwin and evolutionary thought," *Evolution and Anthropology: A Centennial Appraisal* (Theo. Gaus' Sons, Brooklyn), 1–10; 1960, "The emergence of evolutionary novelties," in Tax (1960), Vol. 1, 349–380; 1961, "Cause and effect in biology," *Science* 134: 1501–1506; 1963, *Animal Species and Evolution* (Harvard University Press); 1966, Introduction to facsimile edition of Darwin, 1859 (Harvard University Press).

McAtee, W. L., 1932, "Effectiveness in nature of so-called protective adaptations in the animal kingdom, chiefly as illustrated by the food-habits of nearctic birds," *Smithsonian Miscellaneous Collection*, Washington, 85, 7: 1–201; 1937, "Survival of the ordinary," *Quarterly Review of Biology* 12: 47–65.

Medawar, Sir Peter B., 1960, *The Future of Man* (Basic Books).

Moore, Ruth, 1953, *Man, Time, and Fossils* (Knopf); 1956, *The Earth We Live On* (Knopf).

Morris, Desmond, 1967, *The Naked Ape* (McGraw-Hill; paperback by Dell, 1969).

Nicholson, Alexander J., 1960, "The role of population dynamics in natural selection," in Tax (1960), Vol. 1, 477–521.

Olson, Everett C., 1960, "Morphology, paleontology, and evolution," in Tax (1960), Vol. 1, 523–545; 1965, *The Evolution of Life* (Mentor).

Osborn, Henry Fairfield, 1906, "The causes of extinction of Mammalia," *American Naturalist* 40: 769–795, 829–859.

Overhage, Paul, and Karl Rahner, 1963, *Das Problem der Hominisation* (Herder Verlag, Freiburg i.B., Germany).

Petrunkevitch, Alexander, 1952, "Macroevolution and the fossil record of Arachnida," *American Scientist* 40: 99–122.

Popper, Sir Karl R., 1963, *Conjectures and Refutations* (Routledge and Kegan Paul, London).

Portmann, Adolf, 1964, *New Paths in Biology* (Harper & Row).

Rensch, Bernhard, 1960, *Evolution Above the Species Level* (Columbia University Press).

Rickett, Harold William, 1968, "The changing names of plants," *Garden Journal*, January/February, 12–14.

Sanderson, Ivan T., 1960, "Riddle of the frozen giants," *Saturday Evening Post*, 16 January, 39, 82, 83.

Shaw, George Bernard, 1921, *Back to Methuselah* (Penguin paperback).

Simpson, George Gaylord, 1944, *Tempo and Mode in Evolution* (Columbia University Press); 1949, *The Meaning of Evolution* (Yale University Press; rev. ed. 1967); 1953, *The Major Features of Evolution* (Columbia University Press; paperback by Simon & Schuster, 1967); 1964, *This View of Life* (Harcourt, Brace & World); 1969, *Biology and Man* (Harcourt, Brace & World).

Smith, John Maynard, 1958, *The Theory of Evolution* (Penguin).

Standen, Anthony, 1950, *Science Is a Sacred Cow* (Dutton).

Stebbins, G. Ledyard, Jr., 1950, *Variation and Evolution in Plants* (Columbia University Press).

Tax, Sol, ed., 1960, *Evolution After Darwin* (University of Chicago Press). This comprises three volumes entitled respectively, *The Evolution of Life, The Evolution of Man,* and *The Issues of Evolution.*

Tinbergen, N., 1954, "The origin and evolution of courtship and threat display," in Huxley, Hardy, and Ford, 1–71.

Vavilov, N. I., 1951, *The Origin, Variation, Immunity, and Breeding of Cultivated Plants* (Chronica Botanica).

Velikovsky, Immanuel, 1955, *Earth in Upheaval* (Doubleday).

Waddington, C. H., 1960, "Evolutionary adaptation," in Tax (1960), Vol. 1, 381–402.

Whitley, D. G., 1910, "The ivory islands in the Arctic Ocean," *Philosophical Society of Great Britain, Journal of Transactions,* 42: 35–56.

Whittaker, R. H., and P. P. Feeny, 1971, "Allelochemics: chemical interactions between species," *Science* 171: 26 February, 757–770.

Williams, George C., 1966, *Adaptation and Natural Selection* (Princeton University Press).

Wright, Sewall, 1960, "Physiological genetics, ecology of populations, and natural selection," in Tax (1960), Vol. 1, 429–475.

Zuckerman, Sir Solly, 1954, "Correlation of change in the evolution of the higher primates," in Huxley, Hardy, and Ford, 301.

Index